SpringerBriefs in Environmental Science

SpringerBriefs in Environmental Science present concise summaries of cutting-edge research and practical applications across a wide spectrum of environmental fields, with fast turnaround time to publication. Featuring compact volumes of 50 to 125 pages, the series covers a range of content from professional to academic. Monographs of new material are considered for the SpringerBriefs in Environmental Science series.

Typical topics might include: a timely report of state-of-the-art analytical techniques, a bridge between new research results, as published in journal articles and a contextual literature review, a snapshot of a hot or emerging topic, an in-depth case study or technical example, a presentation of core concepts that students must understand in order to make independent contributions, best practices or protocols to be followed, a series of short case studies/debates highlighting a specific angle.

SpringerBriefs in Environmental Science allow authors to present their ideas and readers to absorb them with minimal time investment. Both solicited and unsolicited manuscripts are considered for publication.

More information about this series at http://www.springer.com/series/8868

Ulrike Pröbstl-Haider · Monika Brom
Claudia Dorsch · Alexandra Jiricka-Pürrer

Environmental Management in Ski Areas

Procedure—Requirements—Exemplary Solutions

Springer

Ulrike Pröbstl-Haider
Department of Landscape,
Spatial and Infrastructure Sciences,
Institute of Landscape Development,
Recreation and Conservation Planning
(ILEN)
University of Natural Resources and Life
Sciences
Vienna
Austria

Monika Brom
Department of Sustainable Development
Environment Agency Austria
Vienna
Austria

Claudia Dorsch
AGL—Arbeitsgruppe für
Landnutzungsplanung
Etting-Polling
Germany

Alexandra Jiricka-Pürrer
Department of Landscape,
Institute of Landscape Development,
Recreation and Conservation Planning
(ILEN)
University of Natural Resources and Life
Sciences
Vienna
Austria

ISSN 2191-5547 ISSN 2191-5555 (electronic)
SpringerBriefs in Environmental Science
ISBN 978-3-319-75060-6 ISBN 978-3-319-75061-3 (eBook)
https://doi.org/10.1007/978-3-319-75061-3

Library of Congress Control Number: 2018930118

Printed on acid-free paper

This Springer imprint is published by the registered company Springer International Publishing AG part of Springer Nature
The registered company address is: Gewerbestrasse 11, 6330 Cham, Switzerland

Foreword

Many ski areas today advertise with seals of approval, titles and awards. Often, however, it is unclear which institutions and criteria are behind these appraisals. Both skiers and ski areas criticise this practise of 'greenwashing', since a cautious use of ecologically and scenically sensitive mountain areas is more important now than ever. An environmental management system that builds on international or European standards ensures 'real' improvements and external evaluation. It relies on the personal responsibility of the providers rather than on regulations imposed by authorities.

Applying such instruments is part of the tradition of the Verwaltungs- und Privat-Bank Aktiengesellschaft (VP Bank) in Liechtenstein. The bank has many years of positive experience with environmental management within its own operation. We therefore chose to support the development of an environmental management especially for ski areas with the creation of the pro natura–pro ski foundation.

The present publication is based on the book 'Auditing in Skigebieten' by Ulrike Pröbstl, Ralf Roth, Heiner Schlegel and Rudolf Straub (2003), which was supported by the foundation. New insights and experiences in practice as well as a change of the legal situation call for a revised edition. As president of the foundation committee, I am delighted that we were again able to find such a competent team of authors for this new edition. The book will contribute significantly towards a compatible cooperation and a cautious handling of the mountain area in and around ski areas.

Hans Brunhart
Princely Counsel, President of the Foundation
Committee pro natura–pro ski

Acknowledgements

Worldwide, environmental management systems make valuable contributions towards sustainable development. The field of application for the European Eco-Management und Audit Scheme (EMAS) has increased over the past years and now ranges from wood processing over hospitals to protected areas (see footnote[1]). With this publication, we would like to contribute to increasing the number of certified ski areas and to raise awareness in general for environmentally friendly and resource sparing management in winter sports areas.

This publication would not have been realised without the support of a number of partners. Particular thanks, in this respect, go to the pro natura–pro ski foundation and to the Austrian provinces of Lower Austria and Salzburg for their financial support. The strategic cooperation between the Environment Agency Austria (Umweltbundesamt) and the University of Natural Resources and Life Sciences, which has greatly benefited both institutions, also made this joint publication possible.

We are grateful to experts of the environmental committee of the International Organisation for Transportation by Rope (O.I.T.A.F.) for their practical approach and consideration of current developments, and to partners with practical experience, such as the Arbeitsgruppe für Landnutzungsplanung (Polling/Bavaria), Naturraum-Management Steinwender (St. Veit im Pongau/Salzburg), DI Fritz Pichler and the consulting firm Klenkhart (Absam/Tyrol), as well as Lloyd's Register (Vienna). Thanks are also due to various ski areas that provided image material as well as exemplary solutions and experiences. With their commitment and leadership, they contribute greatly to spreading the idea.

[1]For detailed information on EMAS by the European Commission, see http://ec.europa.eu/environment/emas/index_en.htm.

Special thanks also go to DI Verena Melzer and DI Claudia Hödl for their meticulous collection and preparation of illustrative material and to Benjamin Seaman for the English translation.

Ulrike Pröbstl-Haider
Monika Brom
Claudia Dorsch
Alexandra Jiricka-Pürrer

Contents

List of Figures

List of Tables

Chapter 1
Introduction

Abstract Environmental management is getting more and more important for destination choice, green marketing and sustainable tourism offers. In the past, environmental measures were primarily imposed by conservation authorities. Nowadays, an increasing number of organizations choose to take a more active approach and address environmental issues in their management and marketing schemes. The introductory chapter highlights this voluntary application of environmental management systems by ski resorts and ski areas as a crucial opportunity for sustainable development.

Keywords Destination choice · Green profile · Environmental management

The choice of ski areas and winter holiday destinations has increased steadily over the past years, especially in Eastern Europe and in Turkey. North American ski areas are also increasingly courting European skiers, enticing them with "powder", "heliskiing" and untouched "hinterland".

It will become more important in future, therefore, for ski areas to distinguish themselves from other regions and to develop a unique profile. Since the sports possibilities and snow reliability are comparable in many destinations, other aspects need to be included in destinations' outward presentation.

An increasingly important argument in this respect is the environmental compatibility of the business and its so-called "green profile". Therefore, the issue of environmental management in ski areas is also gaining increased attention in the media (cf. Fig. 1.1).

Forward-thinking regions and winter sports areas apply these measures:

- Compatible piste management creates species-rich mountain meadows, which by far exceed intensively used meadows regarding biodiversity, and even save maintenance cost.
- Infrastructure for snowmaking and water reservoirs are also used for energy production.
- Contracts with railways and municipal bus services reduce the environmental impact of reaching the destination and benefit environmentally aware guests.

U. Pröbstl-Haider et al., *Environmental Management in Ski Areas*, SpringerBriefs in Environmental Science, https://doi.org/10.1007/978-3-319-75061-3_1

Fig. 1.1 Television recording by the ZDF for the ski audit of Planai-Hochwurzen cableways in Schladming (*Photo* Ulrike Pröbstl-Haider)

- An environmental management system informs and trains personnel and motivates them to become involved for the environment in their respective fields.

Of course the question is, how many customers can be attracted by a green profile and environmental involvement.

The image described above does not appeal to all skiers. According to our results, it appeals foremost to the quality-oriented tourist, who wishes to combine sports, recreation and an experience of nature in winter. Regarding regional value added, these customers are particularly important. Current surveys show that the proportion of environmentally sensitised customers is increasing, particularly in the Alpine region. A survey among 1165 winter sports tourists (Pröbstl et al. 2011) in four large ski areas in Austria (Lech am Arlberg, Schladming, Zell am See and Silvretta Montafon) showed that 44% would prefer a ski area that produces its own energy (46% were uncertain and 10% did not consider this aspect important). But it is not just Europe where a change in thinking is taking place. Also in North America, large ski areas like Whistler in Canada are placing more and more importance on environmental aspects.[1]

In Europe, environmental management systems (EMAS and ISO 14001) have established themselves as important instruments for sustainable development. Decisive arguments for participating in these systems are resource savings, improvement of legal security including environmental liability, and motivation of personnel.

After their success and wide acceptance in production, these instruments are now also gradually becoming more widespread in tourism.

A convincing aspect from the point of view of the businesses continues to be that these are not measures imposed by an authority, but are rather voluntary, market economy-oriented measures of environmental precaution. Extent of measures, as well as focus and orientation are largely within the responsibility of the business. The extensive range of possible applications creates an incentive for both strongly and less strongly impacted areas.

The external certification and evaluation provide a level of credibility that cannot be reached by the many self-proclaimed awards and seals of approval. The EMAS evaluation or the ISO-certificate are therefore easy to distinguish from "greenwashing" via seals of approval.

Sustainable management in a winter sports business must be a matter of course. The way a forward-thinking management conveys its achievements for nature and the environment needs to become more proactive, provocative and exciting.

We wish to help spread these ideas with this book. Checklists with a close reference to practical application in each of the fields will assist in this respect.

Reference

Pröbstl U, Jiricka A, Hindinger F (2011) Renewable energy in winter sports destinations—desired, ignored or rejected? In: Borsdorf A, Stötter J, Veuillet E (eds) Managing Alpine Future II "Inspire and drive sustainable mountain regions", Proceedings of the Innsbruck conference, 21–23 November 2011

[1]Further details can be found under the header "environment" on the homepage of Whistler Blackcomb: http://www.whistlerblackcomb.com/environment/index.aspx.

Chapter 2
Aims and Basic Principles of the Handbook

Abstract This chapter introduces the main international standards of environmental management systems and their related certification, illustrating basic principles and rules. Furthermore, the main goals of this handbook are described such as identifying the specific challenges for ski resorts, the particular methodological adaptations that need to me made and the prerequisites for a successful certification.

Keywords ISO 14001 · EMAS · International standards

2.1 Aims

The handbook aims to encourage cableway operators and ski area managers to introduce environmental management systems (EMS), and provides assistance in this respect.

It should illustrate the relevant aspects and the set of methods that can be applied. Based on previous research and development projects, the gained experiences and insights are summarised in this handbook. Based on the already developed "Ski-Audit" instrument, insights gained from integrated, transnational assessment procedures are conveyed using best practice examples from the Alpine countries.

Based on the principles of an environmental management system prescribed by international rules and standards (EMAS, ISO 14001), the aim was to provide preconditions, data and recommendations for an efficient implementation. Due to the multitude of European directives (e.g. Environmental Liability Directive, Birds Directive and Habitats Directive), environmental management in ski areas has gained importance significantly, especially in the Alpine area and other European winter sports destinations. This book dedicates itself to these new challenges, but also to the tasks required to further develop existing ski areas.

U. Pröbstl-Haider et al., *Environmental Management in Ski Areas*, SpringerBriefs in Environmental Science, https://doi.org/10.1007/978-3-319-75061-3_2

We hope to contribute to a raising of standards worldwide, and to a process of learning from one another within the industry.

2.2 Basic Principles (Requirements of ISO 14001 and EMAS)

When speaking of environmental management, we refer here to measures prescribed by two sets of rules and standards, which both have the same goals but differ in their histories of development, their areas of application and also slightly in their requirements:

- The international standard for environmental management ISO 14001 sets the criteria for an environmental management system and is part of a whole family of standards. It was first developed in 1996 as 14001:1996. The current version of ISO 14001 took effect in September 2015 as ISO 14001:2015.
- EMAS stands for "Eco-Management and Audit Scheme". It is a voluntary system in which businesses but also other organisations and institutions of the EU member states can participate. The regulation is in effect since April 1995 and has already been revised twice. EMAS III is in effect since 11 January 2010 (Regulation (EC) No. 1221/2009). With this revision, it became possible for organisations from outside of the EU to participate. General aim of the regulation (as with ISO 14001) is to promote a continuous improvement of organisations' environmental performance.

The table below gives an overview of the areas in which the two instruments differ. This book will largely highlight EMAS since this is the more far-reaching instrument (see Table 2.1).

A number of businesses make use of the extensive conformities between the requirements of the two systems. They register for both ISO 14001 and EMAS certification, since this does not require much greater internal effort (e.g. Lech am Arlberg from 1999 to 2006). An integrated management system in combination with safety-related aspects is another obvious option. Businesses and organisations often wish to manage various areas "systematically" and can thereby regard quality, environmental protection, work safety and health and safety in an integrated manner.

Simultaneous application of ISO 14001 and EMAS is mostly chosen in cases where the two different systems should appeal to different target groups (e.g. customers from outside the EU) or the extensive level of examination should be documented and outwardly communicated.

Table 2.1 Comparison of features and requirements of ISO 14001 and EMAS

Features	ISO 14001	EMAS	Differences
Decision on implementation and definition of area of application	+	+	
Definition of environmental policy	+	+	
Procedure for assessing, monitoring, documenting and implementing measures relevant to significant environmental aspects	+	++	In the case of EMAS: consideration of direct and indirect environmental aspects, completion of the first environmental review
Procedure for assessing legal requirements	+	+	
Compliance with requirements of environmental law	+	++	In the case of EMAS: also an assessment of compliance with relevant regulatory requirements by environmental verifiers
Improvement of environmental performance	+	++	More emphasised by EMAS, e.g. through a mandatory selection of core indicators
Measures for realisation and implementation in the organisation	+	+	
Training, awareness raising and participation	++	++	In the case of ISO 14001: particular focus on environmental training In the case of EMAS: particular focus on participation of personnel and flow of information Long-term, process-oriented participation
Communication	+	++	In the case of EMAS: open dialogue with interested parties, environmental statement is publicly available, EMAS-logo
Documentation, internal observation	+	+	

Overall, this handbook should provide:

- concrete assistance in establishing and implementing an environmental management system in a ski area based on conveyed experience and illustration of measures in case examples.
- a clear methodology—particularly where it comes to landscape-related content, which can be hard to grasp.
- the basis for a technical standard for successful certification/evaluation.

Chapter 3
Development of Environmental Management Systems in Ski Areas

Abstract This chapter presents the crucial auditing cycle which is the basis for any eco-management and auditing scheme as introduced by the European Commission. Against these basic principles and their historic development, the main advantages and benefits of a participation in the EMAS system are reported. Finally, the definition and often challenging spatial delimitation of a ski area are described.

Keywords Audit cycle · Risk reduction · Costs · Competitiveness
Organisational benefits · Definition of ski areas

3.1 Basic Principles and Requirements

The Eco-Management and Audit Scheme (EMAS) was founded in 1993 by the European Commission as a voluntary instrument for systematic environmental protection. It is a voluntary system in which businesses, but also other organisations and institutions from all economic sectors inside or outside of the European Union, can participate. The aim of this system is to promote continuous improvement of organisations' environmental protection.

As a future-oriented environmental management system, EMAS can help businesses improve their capacity for innovation, reduce their environmental impact and costs, and strengthen their credibility.

A special characteristic of the EMAS regulation, thanks to the voluntary participation of businesses, is that it is based on businesses' own initiative towards environmental protection and their control thereof. This is achieved through the integration of environmental matters in all fields of business policy. Article 1 of the regulation states that

U. Pröbstl-Haider et al., *Environmental Management in Ski Areas*, SpringerBriefs in Environmental Science, https://doi.org/10.1007/978-3-319-75061-3_3

> the objective of EMAS, as an important instrument of the Sustainable Consumption and Production and Sustainable Industrial Policy Action Plan, is to promote continuous improvements in the environmental performance of organisations by the establishment and implementation of environmental management systems by organisations, the systematic, objective and periodic evaluation of the performance of such systems, the provision of information on environmental performance, an open dialogue with the public and other interested parties and the active involvement of employees in organisations and appropriate training.

No concrete details as to the form of environmental policy, programme or management are made by the regulation, nor are any material parameters such as thresholds or environmental standards defined, to which businesses would be bound beyond the legal requirements. This would hardly be possible, especially in light of the diversity of businesses that can participate in the EMAS system. Thus it is up to the individual business to analyse its environmental impact responsibly and to take steps and measures in accordance with its self-set goals. This is significant because it allows businesses to set themselves goals that are timed to match economic requirements and that are adjusted to their existing environmental protection measures. At the same time, the business can satisfy the regulation's call for continued improvement by steadily increasing its own requirements for optimisation of environmental performance with each cycle (cf. Fig. 3.1).

The last amendment in 2009 saw the addition of global applicability (EMAS Global), increased consideration of the interests of small and medium-sized enterprises (SME), as well as standardised environmental indicators to show improvement of performance (UGA 2014).

Environmental policy, as the roof encompassing all environmental activities, is usually the first step of the procedure. Basic environmental guidelines are formulated here, and are integrated into corporate policy. This ensures that environmental and economic corporate goals are at one level and are pursued equally.

The centrepiece of the environmental management system is the environmental review. It constitutes a first appraisal of an organisation's environmental performance and impact. Aside from compiling all of the organisation's environmentally relevant data, the source data is also analysed and evaluated to determine weak points. The environmental review further includes an assessment of compliance with environmental regulatory requirements. The environmental programme is derived from the results of this (first) environmental review. The environmental programme contains all targets and measures aimed at overcoming the weak points, as well as specific deadlines. It also defines individual persons responsible and resources available for implementation of the measures. In the next step, implementation of the measures from the environmental programme begins.

All structures for the implementation of the organisation's internal environmental protection are defined in the environmental management system as part of the overall management system. Thus, the environmental management system comprises personnel-related organisation structures and responsibilities as well as process control. It also ensures documentation of all environmental activities.

The organisation's overall environmental activities are to be presented to the public in regular environmental statements. The environmental statement thus constitutes

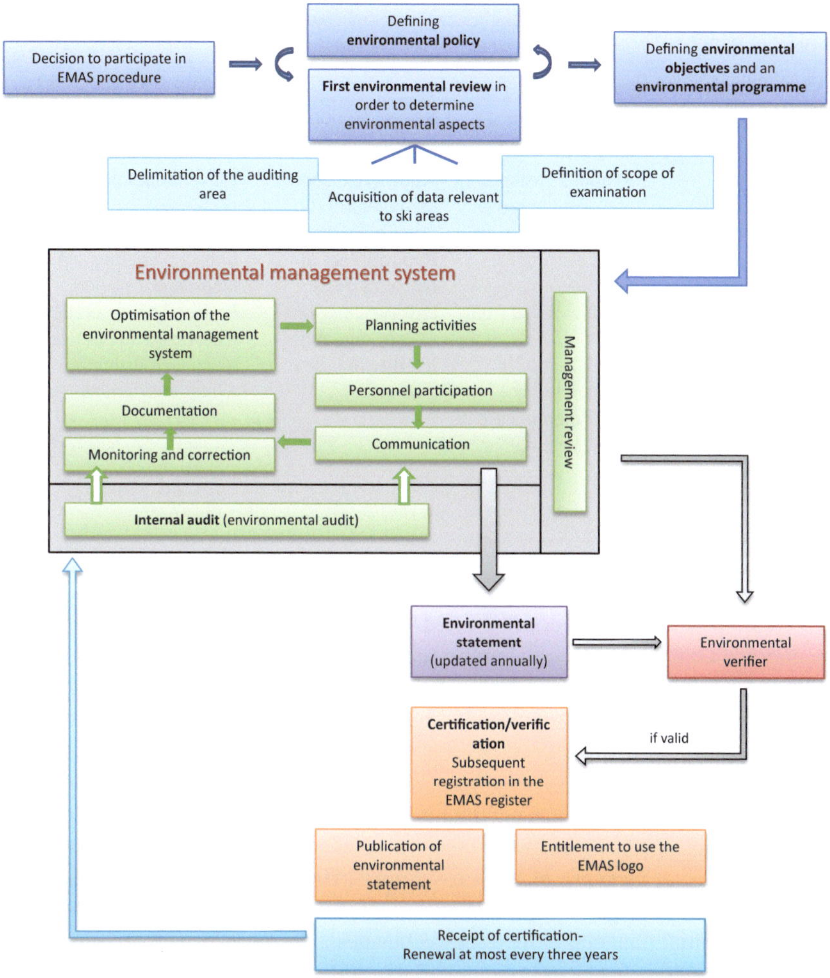

Fig. 3.1 Procedure according to the EMAS regulation (Schublach 2014: 47)

a gateway to the extra-organisational level and ensures communication with the stakeholder groups. It should therefore be kept simple and easy to understand, and should summarise the environmental performance in brief form.

Verification of the organisation is carried out based on the environmental statement and the environmental management system. The process involves an independent, external environmental verifier who assesses whether an organisation's environmental policy, (first) environmental review, environmental programme, environmental management and environmental statement comply with the requirements of the EMAS regulation. If all requirements are met, the organisation can take part in

Table 3.1 Potential benefits for ski enterprises of participating in the EMAS system (Pröbstl et al. 2003)

Potential benefits for the ski enterprise			
Competitiveness	Cost reduction	Risk reduction	Improvement of organisation
Appeal to an environmentally sensitive circle of customers Environmental commitment as a key qualification Improved mutual trust with conservation authorities Easier candidacy for major winter sports events	Reduced costs for preparing approval documents Lower insurance rates Cheaper loans from banks Prevention of cost-intensive restoration in case of large-scale damages	Knowledge about the potential for large-scale damages Reduced risk of damages due to erosion Clarity regarding liability in case of damages	Improved knowledge of personnel Detailed knowledge about impact on nature and the landscape Reduced bureaucratic effort Knowledge about the relative contributions of winter and summer tourism

the EMAS system and be added to the EMAS register. Following registration through the national competent body, the organisation is entitled to use the EMAS logo.

In order to maintain its eligibility in the long-term, the organisation is required to repeat the procedure in regular intervals of at most three years. SME with low environmental impact are allowed to extend the audit cycle to four years.

From the point of view of an organisation, there are a number of arguments in favour of introducing an environmental management system (cf. Table 3.1).

A significant factor in relation to the introduction of an environmental management system is an **increase of business competitiveness**. This is largely thanks to an image boost through an environment-oriented corporate policy. Even if EMAS involves an initial financial burden for the business, the **exposure of savings potentials** such as energy, water or basic materials can lead to cost reduction.

The (first) environmental review uncovers weak points that are then addressed by the business' environmental policy and the environmental programme. Compulsory incident and emergency preparedness also results in **risk reduction**. Establishing an **information system for all environmentally relevant corporate activities** assists efficient internal controlling. This involves breaking down internal processes into the manageable components planning, implementation, monitoring and correction (Landesanstalt für Umweltschutz Baden-Württemberg 1998). Risk management and the information system have a positive effect on legal security and insurance obligations.

Establishing different process components at all levels of the organisation improves **communication with and among personnel**. Integration of environmentally oriented principles and their concrete realisation boosts employees' motivation and contributes significantly to their identification with the business.

Assessing all of the business' environmental impacts in the first environmental review, as well as how they are handled in environmental policy and the environmental programme, constitutes the **basis for sustainable economic activity**. Furthermore, through **increased transparency regarding environmental impacts** and their publication in the environmental statement, a significant improvement of the business' external perception is possible.

EMAS also allows registration of larger and more complex units, such as associations of businesses, supply chains, shopping centres, etc. This cluster approach is also interesting for ski areas, because these usually consist of several different businesses that could thereby register under a single EMAS registration number.

The following options of forming a cluster in cableway-related tourism make good sense:

- A cluster with other cableway businesses, i.e. in cases where a number of installations are active in one resort or on one mountain and the ski area is already run collectively (shared ski passes, piste grooming and preparation, etc.) (close functional relationship).
- A cluster with other businesses that are closely connected to the cableway business. This can include ski rentals, sports dealers, restaurants and even accommodation (broad functional relationship).
- A geographically defined cluster around a larger unit of businesses that are not functionally related, e.g. a nature park. New means of cooperation and efficient management make good sense in this respect.

3.2 Special Requirements in Ski Areas

In order to establish an environmental management system in ski areas, an own set of methods and a review framework need to be developed. On the one hand, this is because EMAS and ISO 14001, with their wide range of possible applications, allow for considerable freedom. On the other hand, this means that each business (or each business type) has to find its own specific way to integrate operative aspects with ecological and landscape-related aspects in a responsible manner. This requires an independent methodological design.

The methodological elements described in the following can also be used for ecological improvements irrespective of environmental management procedures (ISO 14001 or EMAS) (Pröbstl et al. 2003).

The specific challenges that suggest a separate approach for ski areas can be summarised as:

- Combination of high standards regarding passenger transportation and system safety
- Influence of weather conditions on operation and usage of installations
- Safety in the mountains and during activities
- High resource effort for operation
- Great sensitivity of the affected natural habitats due to the high altitude (e.g. vegetation, water, soil, wild animals).

Table 3.2 Definition of a ski area as the area in a landscape influenced by skiing activities (Pröbstl et al. 2003: 42)

Landscape	
Ski area	
Area used directly by skiing activities	Area influenced by skiing (impacted area)
Ski pistes, i.e. usually groomed slopes signposted by the operator and which the operator is required to secure **Routes**, i.e. ungroomed, marked slopes that are merely secured against avalanches and are not controlled or cleared of obstacles **Alternative routes**, i.e. unmarked and unsecured slopes that skiers use at their own risk **Infrastructure**, lift installations, reservoirs for snowmaking, restaurants, etc.	**Areas affected by noise**, that surround the used area and have an impact on adjacent habitats **Piste-forest edges**, that can be affected by skiing activities **Other knock-on effects**, emanating from pistes and alternative routes, such as water discharge from pistes, snow gliding or avalanche blasting

On top of this there are the matters of waste generation and disposal and supply of drinking water, both of which are difficult in mountainous areas.

Above all, the challenges include the geographic delimitation of the ski area, identification and overview of relevant ski area data, and examination of thematic and geographic priorities. For this step in particular, cooperation with external offices already in the preparatory phase is advisable. This is handled differently by different businesses. The ski area Zell am See has installed a permanent ecological advisory board, whose duties include consulting services regarding the established environmental management system. The ski area Kaprun has entered a permanent cooperation with an advisory office for environmental issues. Other ski areas like Lech/Zürs am Arlberg draw on the advice of technical offices when and as needed. In the beginning phase, it is important to focus especially on those areas where internal competence is missing or lacking.

The geographic delimitation should include all areas that are influenced by skiing activities. This project area is defined as the "ski area" (Pröbstl 2001) (cf. Table 3.2).

This comprises not only all sports facilities (lifts, snowmaking and floodlight systems, etc.) and the pistes themselves, but also adjacent areas and off-piste terrain used by skiers.

A number of studies in ski areas (Pröbstl 2001; Dietmann and Kohler 2005) have shown that the project area must be adequately defined. Exclusion of alternative off-piste or lesser frequented routes, for example, is not justifiable from a technical point of view, because valuable habitat for wild animals could be disturbed as a consequence of the operation. Broad delimitation is also important when it comes to management, especially where neighbouring forms of use (hunting, forestry, nature conservation) could be impacted.

Such delimitation is also strongly recommended for the establishment of an environmental management system. Many ecological questions regarding e.g. water balance and fauna, but also recreation, can only be properly answered if considered on a sufficiently large scale (Pröbstl et al. 2003: 41).

Ideally, concrete delimitation is undertaken using topographic maps with altitudinal information or an appropriate digital terrain model (DTM). The delimitation should consider land use in both winter and summer.

The project area is mostly delimited by natural structures such as streams and rivers or mountain ridges. In the case of large, geographically divided ski areas, the project area can also consist of a number of geographically disjunct sub-areas.

Aside from the geographic delimitation with the help of experts, a thematic delimitation is also necessary. In this case, legal aspects, safety issues (Seilbahnen Schweiz 2011) and protection against avalanches and other natural hazards are of particular relevance (cf. BMVIT 2011). Here, again, expert assistance could be advisable in the first stages.

References

Bundesministerium für Verkehr, Innovation und Technologie (BMVIT) (2011) Erlass betreffend den Lawinenschutz im Bereich von Seilbahnen (Lawinenerlass 2011), Vienna, p 8 (in German)

Dietmann T, Kohler U (2005) Skipistenuntersuchung Bayern: Landschaftsökologische Untersuchungen in den bayerischen Skigebieten. Bayerisches Landesamt für Umwelt, Augsburg (in German)

Landesanstalt für Umweltschutz Baden-württemberg (1998) Umweltmanagement für kommunale Verwaltungen. Leitfaden zur Anwendung der Öko-Audit-Verordnung, Karlsruhe, p 62 (in German)

Pröbstl U (2001) Skigebiete in den Bayerischen Alpen. Ergebnisse einer ökologischen Studie. DSV-Umweltreihe, Band 7. Stöppel-Verlag KG, Weilheim. (in German)

Pröbstl U, Roth R, Schlegel H, Straub R (2003) Auditing in Skigebieten. Freiburg, Vaduz (in German)

Schublach F (2014) GIS-unterstütztes Entscheidungstool für Umweltaudits in Schigebieten am Beispiel von Niederösterreich, Masterarbeit an der Universität für Bodenkultur Wien, p 188 (in German)

Seilbahnen Schweiz (2011) Checkliste Verkehrssicherungspflicht für Sommeraktivitäten, Arbeitsgruppe Sommeraktivitäten von Seilbahnen Schweiz, Bern, p 38 (in German)

Umweltgutachterausschuss (UGA) (2014) EMAS—Das glaubwürdige Umweltmanagementsystem. (in German)

Chapter 4
Establishing the Environmental Management System in Ski Areas

Abstract This chapter covers the whole content of the environmental review and the related procedure for obtaining the required information. In this context, an overview on the legal requirements and a legal database are crucial components of the review which should also include safety aspects and emergency planning. The review gives an assessment of the current state encompassing all relevant environmental aspects such as geology and soils, climate conditions, vegetation, fauna, water, immissions, waste and the use of resources. To assist with the assessment and evaluation of these different environmental aspects, related checklists are provided. Finally, personnel participation during the environmental review and the implementation of actions with a clear definition of responsibilities and competencies are explained. The chapter closes with recommendations for communication and documentation.

Keywords Scope of application · Clustering · Environmental policy
Procedural planning · Legal obligations · Safety issues
Environmental review · Danger prevention

4.1 Defining the Scope of Application

The first step in establishing an environmental management system consists of a geographic and thematic definition of the scope of application. The scope can exceed the ski area itself and its associated lift systems. Some ski areas incorporate subsidiary businesses in their operation, belonging to other sectors than cableway operation (e.g. ski rentals, restaurants, souvenir shops and sports dealers). The ski area Bansko in Bulgaria, for example, also includes sports stores and restaurants.

It is important that the organisation (e.g. the business or cluster) introduces an environmental management system in accordance with the requirements of the international standard ISO 14001 or of the EMAS regulation, and documents, implements, maintains and continuously improves it. The system must further contain an explanation of how the requirements are to be fulfilled.

U. Pröbstl-Haider et al., *Environmental Management in Ski Areas*, SpringerBriefs in Environmental Science, https://doi.org/10.1007/978-3-319-75061-3_4

Aside from operational aspects, the geographic area is also crucial for the definition of the scope of application. Regarding delimitation of a ski area, please see Chap. 3 and Table 3.2.

Thanks to the cluster option (Sect. 3.1), new, comprehensive solutions are possible that go beyond an individual business and are focussed on common interests and applications (or areas).

Checklist for scope of application and preparation

- ☐ Is the geographic delimitation of the ski area adequate regarding the activities emanating from the cableway operation?
- ☐ Which additional parts of the operation are included in the review? E.g. restaurants, ski schools, ski rentals, shops?
- ☐ Would the formation of a cluster make good sense?
- ☐ Is there common ground regarding environmentally relevant services or activities, such as waste water treatment or waste disposal shared with other organisations, or regarding outsourced activities like maintenance and service?
- ☐ Has a suitable project team been assembled or are additional external consultants required?

4.2 Environmental Policy and Procedure

4.2.1 Environmental Policy

Environmental policy is the backbone of the entire environmental management process. It is the foundation for all subsequent corporate involvement, addresses the decisive issues and communicates inwards (to personnel) and outwards (to customers and suppliers) that these issues are of the utmost concern to the business' management.

According to the EMAS regulation, **top management** shall define the organisation's environmental policy and ensure that, within the defined scope of its environmental management system, it:

(a) is appropriate to the nature, scale and environmental impacts of its activities, products and services;
(b) includes a commitment to continuous improvement and prevention of pollution;
(c) includes a commitment to comply with applicable legal requirements and with other requirements to which the organisation subscribes and which relate to its environmental aspects;

(d) provides the framework for setting and reviewing environmental objectives and targets;
(e) is documented, implemented and maintained;
(f) is communicated to all persons working for or on behalf of the organisation; and
(g) is available to the public.

The environmental policy is determined by a business' management and all corporate activities are subsequently tested for compliance with the environmental policy. However, the list also shows that considerable conceptual work must precede this initial decision through top management. This is elaborated in the following, pertaining to ski areas.

Preparatory work initially regards the environmental effects. Important environmental effects should be addressed. Important, in this case, means that significant and non-significant effects should be distinguished, and that the environmental policy should be formulated in a way that creates a good framework for the subsequent process without constraining it through too many details.

It is essential for the entire process that the adoption of the environmental policy signals to the whole organisation that cooperation, ideas and input are desired from all employees, and that the process is not merely a "bit of extra homework" for a specific workgroup or individual employees assigned to work on environmental matters.

Obligations regarding improvement of environmental performance and environmental precaution, as well as regarding legal compliance, can be formulated in general terms.

The effort required for the establishment of a management system and for the necessary systematic documentation and monitoring should be pointed out, and resources should be budgeted accordingly.

Internal and external dissemination deserve their own point in the policy. Communication must also involve the public.

The environmental policy as well as its consideration within the business (e.g. in board meetings) and its internal and external communication must convey the message that environmental policy is a matter of the utmost importance in this business. It must be absolutely clear that the environmental policy is a core of the business philosophy, on par with providing high-quality services to customers. It is ideal, therefore, if the business already has a publicised mission statement that either already includes the environmental policy or that can be used as a basis to expand upon the existing objectives.

The environmental policy must comply with the EMAS regulation. Nevertheless, sector and business specific characteristics can and should find expression. In other words, individual formulation of the environmental policy with regard to the respective situation is by all means welcome (cf. Fig. 4.1).

Overall, adopting the environmental policy should be followed by further steps of the environmental management system procedure in a timely manner, because mere adoption of a catalogue of general objectives has little practical effect. Environmental

The green cable car

"The green cable car" describes the efforts of Planai-Hochwurzen-Bahnen's comprehensive, sustainable, resource-efficient and environmentally-oriented company policy.

More than 99% of the slope surfaces are located on third party-owned ground. The Planai-Hochwurzen-Bahnen places great emphasis on a working partnership with agriculture, land owners, forest cooperatives, the local population and institutions and organisations active in the areas of nature and the environment (e.g. the Alpine Association, mountain guides, etc.).

Fig. 4.1 Excerpt from the environmental policy of the Planai-Hochwurzen-Bahnen [Planai-Hochwurzen-Bahnen G.m.b.H. 2017 (The green cable car: http://www.planai.at/de/service/ueber-uns/technik-umwelt#Gruene%20seilbahn)]

policy decreed from "above" is perceived as a compulsory exercise that excites little motivation and has no sustainable effect. It is better to involve the employees in jointly laying the foundations for the environmental policy, implementing it through the introduction of the environmental management, and defining responsibilities and methodological details.

The environmental policy is to be made available to the public. The internet lends itself to this purpose. Frequently, however, the adoption of an environmental policy is also used as an opportunity for a press release aimed at early positioning on the market and a marketing-effective presentation to customers (skiers, tourism partners, etc.). Ideally, participation is also communicated within the business, e.g. during an employee meeting. Not only should the framework of the policy be discussed, but the entire process, including a presentation of key persons within the business as important partners. If assistance through external partners is involved, this would also be an ideal time to introduce them to the business in general. Furthermore, all available media is usually used—from the notice board to the intranet.

Before the environmental policy is communicated externally, it needs to be properly introduced as an internal issue, and the development process to which employees can contribute must be laid out. The environmental policy is all the more effective the better informed the employees are regarding the business philosophy, or the more they were included in its development.

Practice shows that concise, accessible and descriptive formulations should be chosen, when formulating the environmental policy, to maximise acceptance and willingness to cooperate.

A comparison of different ski area environmental policies shows that general formulations and generalised commitments are usually addressed. The following list gives an **example of the contents of an environmental policy for ski areas**:

- **Basic principle**

Environmental policy, in particular the compatibility of our actions with the natural and cultural landscape, is one of the pillars of our corporate activities alongside economic efficiency and social responsibility, and is an integral part of our "corporate philosophy".

- **Scope of efforts**

The business complies with conservation standards and laws and commits to a general reduction of negative environmental impacts due to our economic activities with all available technical, planning and organisational means.

- **Delimitation**

Review, assessment and management of environmental impacts pertain firstly to nature and the landscape. However, the environmental management system also includes technical environmental protection (energy consumption, pollution, noise emission, waste water and waste).

- **Significance within the business**

Environmental protection is a matter of the highest management concern. We therefore promote competence, responsibility and awareness at all levels of the operation through in-formation, training and motivation.

- **Environmental management system**

Our business has introduced an environmental management system. We analyse the environmentally relevant impacts of our operation, define detailed environmental objectives, design an implementation programme and settle responsibilities.

- **Monitoring**

We regularly monitor and assess the compliance of our activities with the environmental objectives. The environmental impacts of our activities are continuously minimised and our corporate environmental performance is constantly improved. We will assess the ecological consequences of new projects, activities, processes and equipment in advance.

- **Business environment**

Wherever possible, we include our suppliers, contractors and customers in our environmental objectives, especially when it comes to obtaining new equipment or contracting projects.

- **Information**

The public is informed about our intentions regarding environmental policy. We provide continuous information about the environmental activities of our business and the results achieved.

- **Adaptation/Development**

We continuously adapt our environmental objectives to the latest environmental insights and to the knowledge we have gained about our ski area.

The environmental policy is part of the environmental statement. It is here that the relation between the general objectives and concrete measures becomes clear and comprehensible.

4.2.2 *Procedural Planning*

Next to the environmental policy, business management must immediately attend to a further important step.

The business must consider procedural steps to approach and implement the tasks described above. This includes organisational aspects but also the allocation of funds and assignment of responsibilities.

As mentioned above, this can also involve the participation of external offices, technical experts or consultants.

In the case of the EMAS regulation, this procedure ensuring that significant environmental aspects are complied with during introduction, implementation and continuation of the environmental management system, is the (first) environmental review. This instrument is described in detail in Sect. 4.4.

Checklist for environmental policy and procedural planning

- ☐ Has the environmental policy been laid down in writing and does it contain overall environmental objectives and principles of action?
- ☐ Was the environmental policy put into effect by top level management or equivalent bodies?
- ☐ Is the scope of application of the environmental policy (spatially and thematically) clear and comprehensible?
- ☐ Does the environmental policy sufficiently address the significant environmental impacts (of the organisation, business or businesses) and the central activities?
- ☐ Is the environmental policy written in a generally accessible manner?
- ☐ Does the environmental policy contain statements regarding compliance with legal environmental regulations and standards?
- ☐ Does the environmental policy contain statements regarding prevention of environmental impacts as well as continuous improvement of environmental performance?
- ☐ Has the personnel been informed about the environmental policy and the objectives of environmental management?
- ☐ Has the environmental policy been made available to the public?
- ☐ Is the environmental policy in line with the general business philosophy and with any existing provisions of higher-level organisations?
- ☐ Have sufficient resources been allocated to implement the environmental management?
- ☐ Have qualified employees been assigned to the procedure and been vested with responsibilities?
- ☐ Are external verifiers required for individual tasks?
- ☐ Has a realistic schedule been established?

4.3 Legal Obligations

4.3.1 Overview of the Relevant Legal Requirements for Ski Areas with Lift Systems

The EMAS regulation places a particular focus on legally compliant operation. This is of particular importance regarding an environmental management system in the winter sports and cableway sector, because a number of areas of law are applicable in this case. This is one of the reasons why the EMAS regulation is recommended in particular in this sector.

According to the EMAS regulation, the organisation shall establish, implement and maintain (a) procedure(s):

(a) to identify and have access to the applicable legal requirements and other requirements to which the organisation subscribes related to its environmental aspects; and
(b) to determine how these requirements apply to its environmental aspects.

Furthermore, the regulation states that the organisation shall ensure that these applicable legal requirements and other requirements to which the organisation subscribes are taken into account in establishing, implementing and maintaining its environmental management system.

Regarding legal compliance, organisations wishing to register with EMAS shall be able to demonstrate that they:

(1) have identified, and know the implications to the organisation of all applicable legal requirements relating to the environment, identified during the environmental review according to Annex I.
(2) provide for legal compliance with environmental legislation, including permits and permit limits; and
(3) have procedures in place that enable the organisation to meet these requirements on an ongoing basis.

Practice shows that the requirements regarding documentation and establishment of a legal database can significantly minimise liability risks, particularly regarding administrative law. Risks regarding civil and criminal law are also significantly reduced through documentation of compliance with the legal provisions.

The diverse and complex subject areas applicable for ski areas illustrate clearly why such attention must be paid to legal aspects in this case. The various legal aspects are addressed in short in the following with examples. All areas of law listed need to be examined in each individual case.

Regarding construction and operation of ski pistes, aspects of contractual law with potential environmental relevance also need to be examined. These include:

Table 4.1 Template for the creation of a legal database

Creation of a legal database—Example						
No.	Legal regulation	In its current version	Text	Interval and date of last review	Responsibility	Status (e.g. reviewed and complied with)

- Questions of ownership/property acquisition
- Lease agreements
- Easements
- Compensation to property owners (in easement contracts).

Lease agreements do not only apply to ski pistes and supply and disposal lines, but also to access roads and water and power lines for snowmaking. Water rights, usage rights for spring catchments and water extraction rights also belong in this category. It must also be considered that legal agreements with property owners can include not only payments but also maintenance obligations, fertilisation, etc.

Regarding installation and operation of lift systems, the contracts apply to the cableway construction (construction firms, system provider, etc.) and respective warranty. Service contracts regarding ongoing operation and energy supply contracts with electricity providers are also important. If energy is fed into the grid by the ski area, then these contracts also need to be considered.

Other legal documents requiring consideration are permits from authorities and the respective requirements and obligations connected to these permits. This includes, for example, obligations and requirements resulting from an environmental impact assessment or accompanying landscape conservation plan, as well as compensation areas and measures, management requirements, etc.

In addition to the permits under conservation law, there are usually also permits required under forestry law (for forest clearing) and water law (for water extraction or introduction).

In some cases the permits also consider the requirements of the Alpine Convention, but this does not usually result in additional obligations.

The obligations connected to permits under conservation, forestry, water and construction law are to be complied with indefinitely. It is therefore a great advantage to compile all legal bases for operation (lease and easement agreements, etc.), as well as all permits and obligations under environmental law in a legal database (cf. Table 4.1).

A greater difficulty is the compilation of legal bases that are not directly connected to administrative decisions and associated obligations, or to agreements under private law.

The business is obliged to determine further relevant legal obligations. In ski areas, these frequently include worker protection provisions, safety regulations and liability issues (e.g. liability disclaimer, indication of FIS rules, etc.).

Regarding autonomous assessment of relevant environmental legal obligations, consideration of European directives plays a key role. These include the Environmental Liability Directive and the Annex IV species of the Habitats Directive.

Considering the particular relevance of these aspects and the fact that they have to be addressed autonomously, important information and assistance for these guidelines is compiled in the following.

4.3.2 Specific Legal Challenges

4.3.2.1 Environmental Liability and Environmental Damages

The term "environmental damage" was coined through the requirement under Community law to implement the Environmental Liability Directive, and is valid in all European Community member states. It pertains to potential impacts on subjects of protection like water, soil, species and natural habitats (Voets 2009; European Union 2013).

The Environmental Liability Directive is not applicable in the case of damages due to causes (events, incidents) before 30 April 2007, nor if the cause of the damage is more than 30 years in the past.

An environmental damage is a damage to protected species and natural habitats. It describes any damage that has significant adverse effects on reaching or maintaining the favourable conservation status of such habitats or species. The significance of such effects is to be assessed with reference to the baseline condition, taking into account the criteria set out in Annex I of the Environmental Liability Directive. In the case of an animal or plant species, these include the number of individuals, their density and the area covered, the species' capacity for propagation, the habitat's capacity for natural regeneration and the significance of the damage (at a local, regional and higher level, etc.).

Furthermore, the term "environmental damage" also pertains to water damages, i.e. any damage that significantly adversely affects the ecological, chemical or quantitative status or the ecological potential of the waters concerned. Operators are liable for the reparation of damages if these are found to be significant and if a causal connection between the damages and the activities of the business or operator can be determined.

It is important to note that this directive does not apply to cases of personal injury, to damage to private property or to any economic loss.

Regarding environmental liability, another new aspect is the so-called "presumption of causation". Operations that intentionally or negligently cause biodiversity damages must remedy those damages in addition to any other environmental liability. If biodiversity damage cannot be fully remedied those liable must take compensatory measures. If a facility is deemed inherently suited to cause a given damage, then the facility is presumed to have actually caused that damage. The burden is on the facility operator to disprove damage causation. Strict standards apply for this procedure.

The significance of environmental management systems has increased because they facilitate such disproval procedures.

Thanks to the Environmental Liability Directive, non-governmental organisations (NGOs) and citizens are now also allowed to inform relevant authorities of any environmental damages (or the imminent danger of environmental damage). They can further take legal action against measures or the omission of measures by such authorities, to ensure in the public interest that those authorities prevent environmental damages with all suitable methods and take all necessary safety measures (European Union 2013: 6).

Specific aspects related to water and biodiversity are briefly addressed in the following:

For issues of water management in relation to environmental liability, the following circumstances are relevant:

- Use of surface or groundwater in quantities required for the purpose of snowmaking, under consideration of third-party rights and the condition of the respective body of water.
- Snowmaking within areas specially protected under water law is only permitted after assessment of each individual case.
- Impact of water applied as snow (and subsequently melted) on vegetation, soil, soil water balance and drainage properties.

Biodiversity in European ski areas is not protected as a whole. Species and habitats are only protected in accordance with the Birds and Habitats Directives in member states of the European Community (Fig. 4.2 shows an example for a protected habitat type on a ski slope). This already significantly narrows the risk of environmental damages occurring (cf. migratory bird species according to Article 4(2) of the Birds Directive, protected bird species in Annex I of the Birds Directive, animal and plant species in Annex II of the Habitats Directive and the list of animal and plant species in need of strict protection referred to in Annex IV of the Habitats Directive).

The purpose of the Environmental Liability Directive is to have a preventative effect and to ensure that significant environmental damages are remedied. Damages are to be remedied by the polluter. The "polluter-pays" principle is thereby further strengthened and promoted in the EU.

The following damages of protected habitats and species would be classified as not significant (European Union 2013: 10):

- Damages of a lesser scale than the natural fluctuation and variability.
- Damages due to a form of management of the affected resource which has been practised in that way over a long period of time.
- Damages of species or habitats which are proven to regenerate quickly without outside influence to the extent that they will reach either the baseline condition or a natural condition equivalent to or better than the baseline condition.

The steps for assessing environmental damages are shown in Fig. 4.3.

Fig. 4.2 Habitat type (according to the Habitats Directive) typical of ski slopes: Hillside spring fen with orchids and cottongrass (*Photo* Rüdiger Urban)

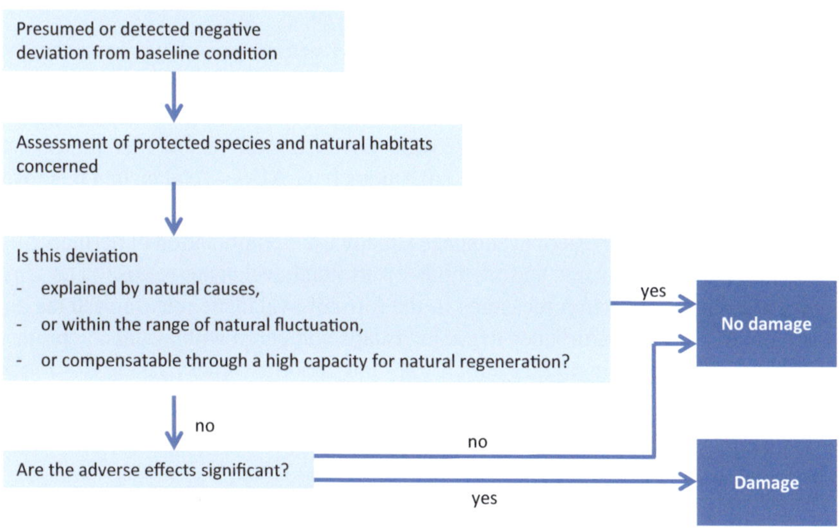

Fig. 4.3 Steps for assessment of environmental damages (following Voets 2009)

The directive does not pertain to damages to protected species and natural habitats wherever the adverse effects result from an act by an operator which was expressly authorised by the relevant authorities in accordance with provisions implementing Article 6(3) and (4) or Article 16 of the Habitats Directive 92/43/EEC, or Article 9 of the Birds Directive 2009/147/EC, or in accordance with equivalent provisions of national law on nature conservation. Such cases include Appropriate Assessments and species conservation assessments (saP) that have determined no impact, as well as administrative decisions.

4.3.2.2 Safety

Avalanche Protection

The issue of avalanche protection is of enormous significance in mountainous areas. Protecting the skiing space from avalanches is already a core element of the approval of a ski area. In parts of a ski area with an avalanche risk, the protective measures are usually a combination of permanently installed technical measures (e.g. barriers) and semi-permanent measures, ranging from closures to artificial avalanche release (bmvit 2011). The operator must ensure facility and operation saftey of cableways and pistes with regard to this issue. This includes examination by experts, e.g. a local avalanche commission, not only of the cableway itself and its associated pistes and access areas to the stations, but also access areas to the cableway (rescue and recovery areas). The measures demanded by the commission must be complied with in this case.

Ski areas that are not safe from avalanches by nature, must have an avalanche safety concept in place, summarising these measures. Diagnosis systems are used in order to judge local conditions, which can depict risk situations based on diverse influencing factors such as topography, snow cover structure, snow metamorphosis and temporal and small-scale climatic conditions (e.g. ADS—Avalanche Diagnosis System) (Keiler et al. 2006).

Most ski areas with a risk of avalanches already use a combination of permanently installed technical measures and methods of artificial avalanche release. The temporary avalanche protection measures in the form of avalanche release systems are much less of a visual disturbance to the landscape compared with avalanche protection barriers, due to their slighter dimensions and unobtrusiveness. Aside from such measures, in some areas it is also possible to restore or improve natural avalanche protection in the long term through reforestation.

Safety in Organised Skiing Spaces

Detailed safety concepts belong to the standards of well-managed businesses worldwide. These include the "ski resorts safety plans" and "ski resorts accident prevention plans" in the USA (CSSSO 2009; Mingo 2000) as well as the title "Geprüftes Skigebiet Deutschland" (certified ski area Germany) awarded by the Stiftung Sicherheit im Skisport (foundation for safety in skiing) (König 2014). Frequently, interdependencies between safety and environmental aspects are not adequately considered. On the one hand, such interdependencies are a result of different legal obligations to

Table 4.2 Legal obligation to maintain safety

Organised skiing space		Free skiing space
Piste	Ski route	Alternative route
Marked	Marked	Unmarked
Graded with sufficient width	No defined width	Not graded
Mostly groomed	Ungroomed	Ungroomed
Controlled	Not controlled	Not controlled
Secured against atypical and alpine hazards	Secured against avalanches	Not secured against alpine hazards

maintain safety between organised skiing space and free skiing space (cf. Table 4.2).

The organised skiing space represents the entirety of ski pistes and/or routes. It does not include alternative routes. Outside of the organised skiing space, it is generally on a skier's personal responsibility to enter the "free skiing space". Furthermore, safety issues are influenced by the perceived danger caused by the environmental conditions and the respective personal responsibility. Regarding the legal obligation to maintain safety on ski slopes, typical and atypical hazards are differentiated. Typical hazards usually include obstacles inherent in the terrain, moguls, individual spots clear of snow and variable snow conditions, but also marking rods and signposts.

From the point of view of environmental management, the business must eliminate potential atypical hazards or give sufficient warning (Dambeck and Wagner 2007). Hazards are considered atypical if a skier cannot be expected to reckon with them, given the appearance and marked difficulty of the piste. These include obstacles such as concrete blocks, ditches, stones or boulders. They are only considered atypical, however, if they are not already visible from a distance (Dambeck and Wagner 2007). National legislations need to be considered because they can vary in these points, also regarding the personal responsibility of the skier.[1]

In the case of legal disputes, judgement is based upon the extent of the hazard, degree of perceptibility, reasonability of security measures and capability of those responsible for ensuring safety, as well as the type of measures (e.g. warnings, protective measures, removal or fencing off). If a business considers a hazard's degree of perceptibility to be low and prefers to remove it rather than rely on warnings or protective measures, then this can have environmental effects. The legal obligation to maintain safety is limited in time to the operating times of the lift system and usually ends with the control run. This aspect is significant in relation to environmental management. Clarity of the piste edge is especially important in this regard. Narrow sections, piste crossings as well as crossings with footpaths or roads, tow lifts and facilities for piste grooming need to be indicated by warning signs if they can constitute a hazard. The legal obligation to maintain safety also pertains to areas away from the piste, e.g. the paths to the parking areas, toilets or transportation facilities.

[1] For Austria, see Manhart and Manhart (2015) and Stabentheiner (2016).

The question of liability is also influenced by publicised codes of conduct (e.g. the FIS rules) and other information that has a positive effect on self-protection. "Too much" signposting can be disadvantageous, however, because it still needs to be effective in varying snow and weather conditions. What is more, extensive signposting can give the subjective impression that no hazards need be observed wherever warning signs were not put up.

Overview of the most important safety aspects in ski areas (winter use):

- Overview maps for orientation for all forms of use, with guidance system, section overviews with difficulty levels and currently closed-off areas
- Information about potential hazards (e.g. avalanches, snow conditions, change of weather)
- Clearly marked piste edges, particularly at crossings, turn-offs and in areas with increased hazards outside of the slopes
- Indication of hazard areas
- Securing of atypical hazard areas
- Securing of piste edges with atypical hazards
- Publication of code of conduct
- Point of contact for security issues within the business
- Publication of emergency facilities (e.g. mountain rescue service).

So-called safety plans, involving consultation by experts during a piste inspection, have often proven valuable to businesses during the environmental management process.[2] Experts in North America (CSSSO 2009) further recommend regular speed measurements.

Regarding use of the area for summer tourism, there are a number of additional aspects that require consideration (Seilbahnen Schweiz 2011).

Overview of the most important safety aspects for summer use of ski areas

- Safety barriers and protection against natural hazards around stations, facilities and supervised infrastructure.
- Delimitation of the secured and monitored area from unprotected areas (overview maps, regulation of control, signposting)
- Approval and conformity with norms of construction, operation and maintenance of infrastructure (e.g. playgrounds, leisure and sports facilities), as well as rules of use and securing of surrounding areas
- Clear designation of tracks and infrastructure requiring mandatory protection measures
- Information regarding difficulty levels of tracks, via ferratas and other infrastructure, and particularly regarding atypical difficulties, hazards and mixed uses (e.g. hiking and mountain biking)
- Protection against seasonal hazards, rockfall and rockslide

[2]In Germany "Prädikat geprüftes Skigebiet Deutschland", Stiftung Sicherheit im Skisport, König (2014).

- Obligatory documentation, requirements regarding inspection and maintenance logs
- Duties regarding information, inspection and protection in conjunction with rental equipment or offers of cooperating third parties.

Checklist for legal obligations

- □ Were procedures introduced to ascertain relevant legal obligations and have them readily available?
- □ Were the legal obligations considered and reviewed during introduction of the environmental management?
- □ Were the following legal matters reviewed in particular?
  - Contract law (e.g. property issues, lease and easement agreements, compensation payments, service contracts)
  - Permits from authorities and administrative decisions (conservation law, construction law, water law, railway law, etc.)
  - Environment and safety (e.g. legal obligation to maintain safety)
  - Environmental liability (environmental damage legislation, water law, soil protection, conservation legislation).
- □ Were the respective regulations adequately communicated to the responsible persons/departments?
- □ Is there a registry of environmental regulatory requirements and administrative decisions for the site?
- □ Are there persons within the business who are tasked with safety matters?
- □ Are safety-relevant issues comprehensively addressed and documented within the business?
- □ Does the business' insurance cover the diverse liability risks?
- □ Is emergency planning or organisation in place?

4.4 Environmental Review

4.4.1 Introduction

The environmental review is an extensive examination of environmental issues and effects. The technical issues relevant to the environmental review were briefly addressed in the previous chapters. It is essential that an extensive analysis of the current state forms the basis for the first environmental review.

A greater effort is only involved for the first environmental review. Later on, the respective issues only need to be updated and the measures implemented accordingly.

This chapter mainly addresses the procedural step itself. More detailed content-related aspects are then covered in the subchapters.

Depending on the size of the business, environmental reviews are updated in regular intervals. The most important steps need to be reviewed by the business itself.

Data updates are also the basis for the internal audits that are required in regular intervals. Generally, involvement of the personnel during the environmental review is necessary, for technical reasons alone. Their involvement should be arranged in any case, because it deepens their environmental awareness and motivation significantly. The environmental review should be designed as an interactive process between employees and technical officials. Those involved should not be burdened with unnecessary theoretical details (e.g. formal courses of action). The main focus should be on a discussion of factual issues.

The first difficult step during the first environmental review is identifying and differentiating all environmental aspects relevant to a ski area, ascertaining potential negative effects of the operation and creating an overview of the applicable environmental regulations. These aspects should be compiled systematically and fully, and evaluated individually.

The first environmental review above all documents the current state and forms the basis for all further steps. In order to match the extent of the review to the respective business and to organise it according to the given situation, the following steps are recommended:

1. Fist, direct and indirect environmental aspects (cf. EMAS Annex I) must be distinguished. Each site or sub-area must be taken into consideration.
   Direct environmental effects in ski areas include snowmaking and piste grooming—core activities of the business that must be backed with according data and figures.
   Indirect environmental effects involve aspects that are only partially within the influence of the business. These include, for example, skiers' means of travel, behaviour of employees or suppliers and service cycle-related aspects.

Annex I of the EMAS regulation offers further examples of direct and indirect environmental aspects.

2. An analysis of the forms of use in the area can assist in detecting the relevant aspects within the direct environmental effects. In analysing forms of use, winter sports, potential summer tourism use, land use (agricultural and forestry use) and potential further services (ski rental, gastronomy) should be considered and analysed separately (cf. Table 4.3).

An alternative helpful step is to choose an approach related to subjects of protection for further data compilation and compilation of potentially relevant environmental aspects (cf. Table 4.3). In addition to the natural subjects of protection, the areas energy (energy efficiency), waste, noise immission and use of resources regularly need to be considered.

Table 4.3 Potential content and available (a) use-related and (b) environment-related data as basis for the environmental management (adapted from Pröbstl et al. 2003)

(a) Data relating to sports and forms of use

Pistes	Lift installations	Snowmaking	Other winter sports infrastructure	Summer tourism	Land use
Delimitation Intensity of use Tournament/mass sports/homologation	Number, type, technical data Theoretical transport capacity Operation figures winter/summer	Area of applied snow Snowmaking machines/course of lines and pipes Reservoir (volume/water extraction) Snowmaking targets/volume	Floodlight systems Funparks, toboggan runs, ski kindergartens, halfpipe, ice skating parks, etc. Events Cross-country ski runs, etc.	Air sports MTB courses Hiking trails Summer toboggan runs Play area Educational trails, etc.	Agricultural use Forestry use Line routes (water use, electricity) Roads, paths, trails Parking areas Restaurants, mountain huts, etc.

(b) Data relating to the environment

Soil	Water	Climate/Air	Vegetation	Fauna
Geological map Soil map Avalanche map Map of slope instability Damage events Terrain intervention Soil protection areas Forest clearances	Bodies of flowing water Bodies of standing water Water protection areas Water quality Water quantities/low water discharge Catchment areas Water use Supply and disposal	Air temperature Precipitation Progression of snowmelt Emissions	Distribution of forest and open areas Characteristic plant communities Occurrence of red-list species Protected areas Damages to the degree of plant cover	Records and territories of red-list species Home ranges of wild animals/game protection areas Habitat structures

Preconditions for the compilation are the complete data, and usually also a transformation or illustration of the information as well as its evaluation. The qualitative and quantitative evaluation of the environmental effects is generally performed according to self-defined criteria. The evaluation must be transparent to the environmental verifier and the general public. Therefore, it often makes sense to provide environmental verifiers and the general public with the processed and evaluated data rather than the collected original data, and to compile a second report with the raw data that can be used (and continuously updated) during the subsequent reviews.

Table 4.4 Overview of the form for data collection

Subject of protection/Form of use			
Data sources and possible contact persons	Basic information	Priority classification	Reason and comments

Information for the environmental statement is extracted from this second report with raw data and the evaluation, and is prepared in an illustrative manner.

In the following, case studies from various ski areas will demonstrate which degree of detail is to be applied in light of the generally rather large landscape sections concerned, and which evaluation makes sense. In the fields of natural resources management, ecology and energy, consulting or contracting experts for stocktaking is recommended. Here, again, it is important to be able to determine a clear scope of expected services and requirements for this task. The following chapters should provide an important contribution in this respect.

The following ski area-specific part of this handbook should ensure that the environmental review is executed as simply as possible, but nevertheless as detailed as necessary. Additional notes are included in the checklists.

The holistic approach proposed here, related to subject of protection and form of use, guarantees complete and clear handling of these issues. On this basis, assistive material was therefore elaborated for each of the subjects of protection and for the usage-related assessment. The general structure of this material has proven useful in the past (Pröbstl et al. 2003). It should facilitate an efficient handling of the process.

In the tables related to subject of protection and form of use, the first column indicates which administrative data is available in order to assess the local situation (cf. Table 4.4). As a study in very different regions of the Alps showed, this is not the same everywhere (Pröbstl et al. 2003). Alternative possibilities are therefore also mentioned. The second column indicates which basic information is derived from the raw administrative data or own surveys. Priority classes indicate the significance of the respective data and their procurement. Hereby, A stands for high priority, B for medium priority and C for lower priority. The last column describes the priority of this data for the ski area-related audit and indicates the fields of work to which the acquired basic data could be applied (cf. Table 4.5).

The following description begins with the natural subjects of protection. The appraisal based on forms of use is described thereafter. Overall, an assessment scale of 1:5000 has proven to be the ideal basis for an environmental management system. Therefore, the most important area data that needs to be collected, such as

- demarcation of ski pistes
- area of applied snow
- plant communities
- construction work carried out
- damages.

Table 4.5 Sources and priority of information (adapted from Pröbstl et al. 2003)

Soil			
Data sources and possible contact persons	Basic information	Priority	Reason and comments
Geological map Soil map Altitude model (terrain structure) Avalanche cadastre Map of slope instability Cableway operators Asking local experts Documentation by authorities Press reports, correspondence (building applications, files) Forestry authorities Water management authorities Conservation authority Municipality	Slope instability, erosivity	A	Important basis for assessing problems and restoration options
	Soil formation Conducted construction measures and terrain grading by extent (surface, partial and total grading), blasting, cutting and filling	B, C A	Natural soil formation is often disturbed in ski areas, therefore limited explanatory power. Nevertheless an important measure that otherwise needs to be inferred at least generally from the local geology
	Cleared areas	A	Loss of forest has a great impact on stability and the risk of erosion, but also on the protection of existing installations and buildings
	Damage events, landslides	A	Conclusions regarding current and recurring problems, problem zones
	Avalanche release systems	A	Regularly disturbed areas due to safety aspects, piste security, etc.
	Avalanche blasting cableways	A, (B)	Area with permanent intervention, impact on wild animals possible

should be assessed and remain permanently available in a geographical information system at this scale.

4.4.2 Geology and Soils

Negative environmental effects related to geology and soils in ski areas mainly concern ground stability and erosion—issues that can lead to damages and liability problems, as well as negative effects on groundwater. The geological and geomorphological conditions of a ski area therefore play an important role, as does its formation history. These factors allow inferences to be drawn regarding

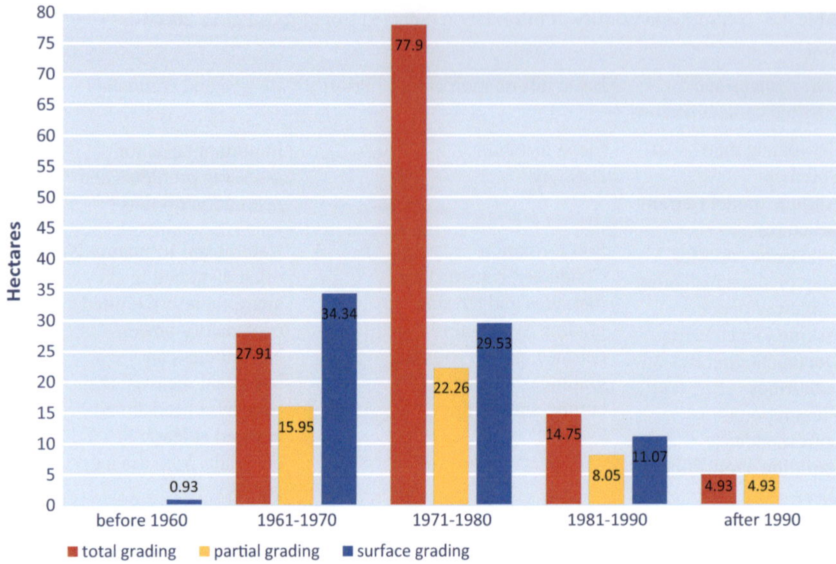

Fig. 4.4 The example of the ski area Hausberg-Kreuzeck-Osterfelder in Garmisch-Partenkirchen (Bavaria), where roughly half of all surfaces were structurally modified, shows that these measures largely took place in the 1970s and that enough time has since past under good conditions to allow regeneration (Pröbstl 2001)

- slope instability or stability
  the sites' regenerative capacity
- growth rate and
- potential consequential effects after construction measures (e.g. grading).

Information on soil formation is important to assess the sites' regenerative capacity and to evaluate potential restoration options. Aspects of soil protection have gained significance in Europe thanks to the Environmental Liability Directive. The protocols of the Alpine Convention—the tourism protocol in particular—also focus on sensitive sites. Regarding ecological stability and naturalness, past construction measures for winter sports, such as grading or clearing, play a decisive role.

Depending on the intensity of the construction measures, graded pistes can exhibit

- altered drainage properties
- altered ground structure as a result of the above
- modified plant cover and
- modified fauna in the soil and in the plant cover.

This means that knock-on effects for different subjects of protection (such as soil, water, vegetation and fauna) can often be inferred from a detailed assessment of construction measures.

Assessment should take clearing, earthmoving, regulation of drainage properties and completed re-cultivation and re-greening measures into account. Construction of infrastructure, lift systems and avalanche control systems (such as blasting cableways) also need to be considered in this context. The year the construction measures were conducted should always be recorded, because this allows conclusions on the regenerative capacity to be drawn (cf. Fig. 4.4). This is possible in particular in cases where the data can be combined with other assessments (e.g. damages). Amongst other things, this is important for the exclusion of liability. Possible data sources and topics that typically need to be included in the assessment are shown in Table 4.5.

4.4.3 Climate Conditions

In order to assess the suitability of an area for winter sports, temperature profiles in the winter months, the number of days of snow and snow depth at various measuring stations offer important basic information. Differentiated information according to altitude is frequently necessary. Avalanche warning service records, regional data from measuring stations or the ski operation's own data e.g. regarding snowmaking (operation journal with temperature records) are all considered here (cf. Pröbstl et al. 2003). The following example from Austria shows how differentiated regional data analysis and consideration can work.

Figure 4.5 shows that, regarding snowmaking, the middle station of the Planai cableway is more of a weak point than the valley station due to less days with low temperatures (below minus 3 °C).

The basic assessment must also consider the conditions during the summer months. An important factor in this respect is the distribution of rainfall, particularly the heavy rainfall events that can increase the risk of erosion in summer. This data is also relevant to the process of snowmaking and its monitoring.

Current environmental management considerations also incorporate model calculations of potential climate change effects. This is a help in planning investments sensibly, scheduling sports tournaments during relatively snow-reliable periods and assessing adaptation measures to climate change. Regarding environmental effects, these results are of great relevance for water and energy demand.

Figure 4.6 compares the number of hours in the past during which technical snowmaking was possible at temperatures below minus 3 °C. A comparison of the periods 1961–1990 and 1988–2002 for this sensitive altitude region (cf. Fig. 4.5) shows that the available time periods have decreased but are still sufficient for operation.

From this basis, scenarios and calculations regarding demand for additional technical snow can be derived. Model calculations considering new technical developments are also possible. The environmental management of the ski area Schmittenhöhe in Zell am See, for example, also assessed the expected snowmaking situation at an assumed temperature of minus 1 °C (Formayer 2013). These contents are taken into account and documented in the environmental review. Possible data sources and topics that typically need to be included in the assessment are shown in Table 4.6.

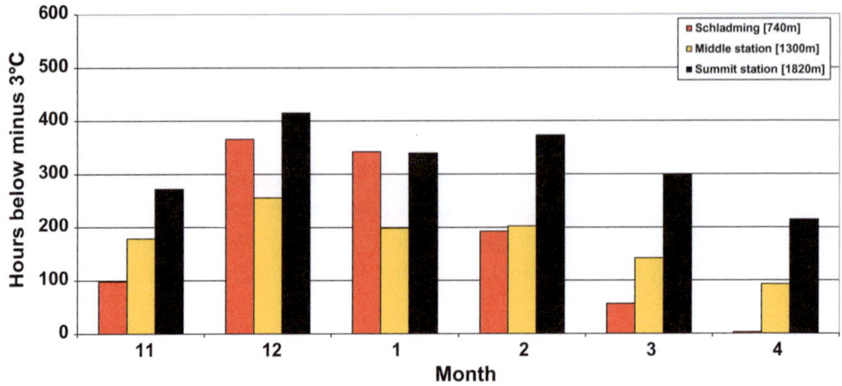

Fig. 4.5 Number of snowmaking hours (i.e. temperatures below minus 3 °C) from 1988–2002 at three altitudes for the Planai cableways in Schladming (Formayer et al. 2007)

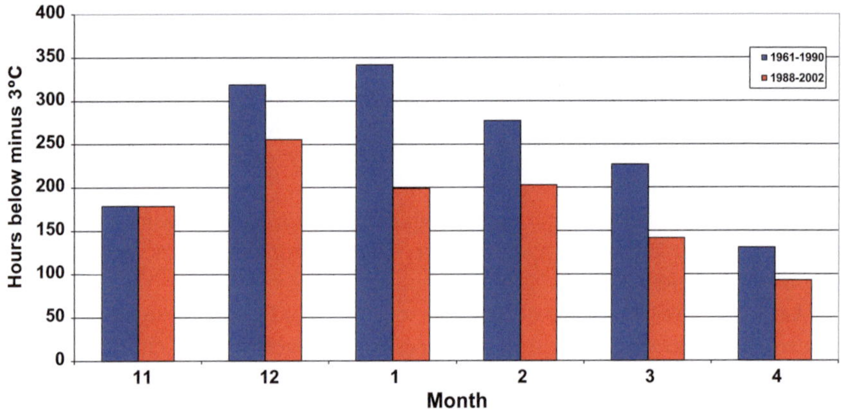

Fig. 4.6 Number of snowmaking hours (i.e. temperatures below minus 3 °C) at the middle station (1300 m) of the Planai cableways in Schladming. Comparison of the periods 1961–1990 and 1988–2002 (Formayer et al. 2007)

4.4.4 Vegetation

The vegetation is an important indicator of the site conditions, the underlying climatic conditions and the intensity of use. Habitat quality for birds, soil fauna and insects can be derived from the naturalness of the vegetation. Furthermore, the vegetation can add to the attraction from a tourism perspective (for example blooming meadows,

Table 4.6 Data assessment for the subject of protection climate (adapted from Pröbstl et al. 2003)

Climate conditions			
Data sources and possible contact persons	Basic information	Priority	Reason and comments
Climate data Operation journal Snow report Climate atlas Climate change scenarios Weather bureau Weather and climate stations Research Institutions	Snow depth by respective altitude and the number of days with snow cover Consideration of hours during which snowmaking is possible	A	Reference values for suitability as a winter sports area
	Distribution of precipitation summer-winter	C	Evaluation of erosion processes in summer Basis for problem assessment and planning of measures
	Temperature (profile in winter in relation to snowmaking)	B	Suitability and applicability of snowmaking
	Progression of snowmelt	A	Explanatory models for observed damages, basis for restoration proposals and demand for snowmaking
	Regionalised climate change scenarios	Depends on the location of the area	Important basic information for long-term environmental management in case of high vulnerability

cf. Fig. 4.7). Regarding direct effects, vegetation plays an important role in the early detection of problem areas and in environmental liability.

A detailed vegetation survey is therefore one of the most important basic surveys during ski area mapping and intended ecological improvement (cf. Figs. 4.8 and 4.9). This is all the more true because studies in German ski areas (Pröbstl 2001; Dietmann and Kohler 2005) and conceptual studies on ecological improvement in Switzerland, Liechtenstein and Austria (Pröbstl et al. 2003) showed that ski pistes cannot be generalised as species-poor, strongly impacted "sports surfaces". On the contrary, these areas harbour an extraordinarily rich structure of small-scale habitats ranging from pioneer communities to high quality areas—a fact which is often also underlined by the occurrence of rare or protected species (Pröbstl 1990; Pröbstl et al. 2003). Figure 4.7 shows the occurrence of rare species in the ski area Lech/Zürs.

The vegetation survey is so important because it not only provides conclusions on naturalness and biodiversity of the ski piste habitat, it also allows an assessment of compatibility of summer and winter use. The vegetation surveys also provide an indication of the self-healing and regenerative capacities (cf. construction measures) of modified sites (Pröbstl 1990, 2006).

Fig. 4.7 The vegetation in many ski areas is diverse and often also includes rare species (image from the ski area Lech/Zürs am Arlberg) (*Photo* Rüdiger Urban)

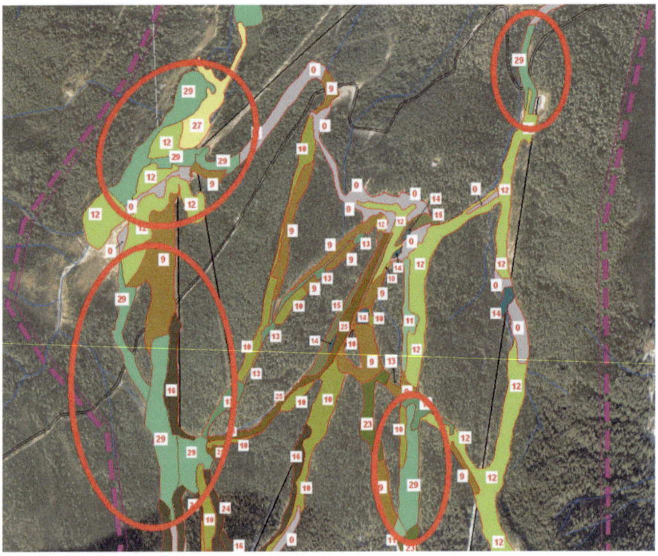

Fig. 4.8 Vegetation surveys show the improvements achieved through piste greening (= green areas circled red) in the ski area Bansko (AGL 2010)

An important part of any assessment of this subject of protection is the survey of damages (cf. the following sections). Additional data sources on vegetation, protected habitats and their priority are given in Table 4.7.

Table 4.7 Data assessment for the subject of protection vegetation (adapted from Pröbstl et al. 2003)

Vegetation			
Data sources and possible contact persons	Basic information	Priority	Reason and comments
Conservation legislation of respective countries or provinces Legislation on environmental liability Protected areas Habitat surveys in open land and forest Details and maps on potential natural vegetation Red-list species of the area Habitat type surveys (especially in line with Habitats Directive) Landscape planning projects Seed mixtures selected for greening Lists and atlases of flora Historic descriptions of the area Authorities NGOs Asking local experts	Separation of forest and open spaces (mountain pines)	A	Significant basis for planning, often difficult to demarcate forest edge and alpine meadows/mountain pines
	Characteristic plant communities	A	Vegetation indicates all abiotic and biotic site factors, as well as impacts
	Red-list species	B	Indication of valuable areas
	Protected sub-areas	A	Basis for planning restoration concepts
	Targets to be considered based on other planning processes (e.g. Natura 2000 management plans)	A	Part of the basis for planning
	Degree of plant cover/damages	A	Starting point for measures, indication of the degree of impact
	Habitat types according to the Habitats Directive	A	Consideration of environmental liability and potential damages to biodiversity Consider protection of surrounding areas in the case of Natura 2000 sites

4.4.4.1 Description of the State of Vegetation

The description of the vegetation should include a description of the potential natural vegetation. This description lists the plant communities that would develop at the site without future human use from the current point of view. This is important to better understand the impact of the human economic activity. However, a rough overview suffices in this case.

Mapping of the terrain with a survey of the current vegetation is generally recommended, and a detailed description and valuation has proven useful (cf. Tables 4.8 and 4.9).

These differentiated descriptions allow subsequent classification of the plant communities to a scale of value representing naturalness. In habitat evaluations (Ellenberg 1963; Plachter 1989; User and Erz 1994), 9-point scales have proven useful

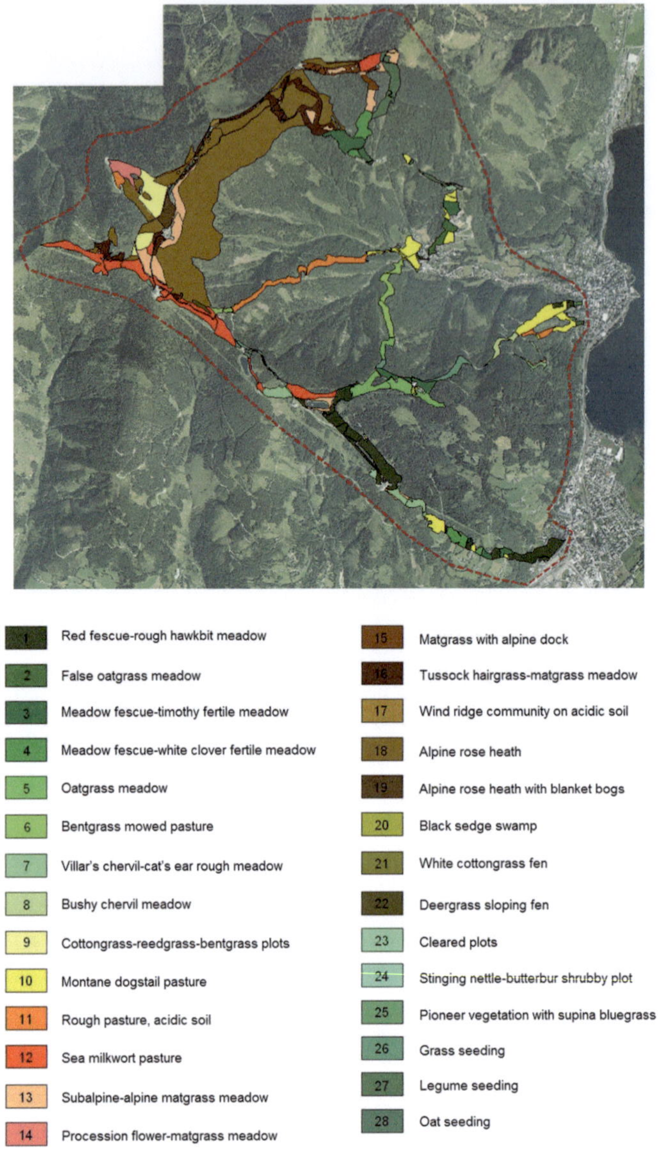

Fig. 4.9 Extract from the vegetation survey in the ski area Schmittenhöhe in Zell am See/Austria (Pröbstl-Haider and Dorsch 2013)

(cf. Ammer and Utschik 1990, amongst others). Even for laypeople, these descriptions illustrate the situation on the ski pistes in an accessible manner. What is more, the 9-point scale can be overlaid or combined with other content in businesses'

Table 4.8 Standardised description of the current vegetation during vegetation surveys (adapted from Pröbstl 2001)

Name of the vegetation unit	No. in the map
Site Occurrence in the survey area, elevation, exposition, further specific site conditions	
Description Proportion of area, degree of cover, description of the community, relation to neighbouring plant communities	
Species array: Species list Characteristic species, dominant species—underlined; Rare species ("red-list species") = protected and/or endangered species: depending on applicable legislation—**bold**	
Significance/threat to the natural balance Significance (worthiness of protection, development trends), acute threats or disturbances, emphasis of rare species ("red-list species") and habitat types relevant to the Habitats Directive, value classification (according to the 9-point scale, cf. Pröbstl 2001), occurrence of species in need of strict protection according to Annex IV of the Habitats Directive	

geographical information systems (cf. the example of the ski area Zell am See in Fig. 4.10).

A methodological approach from ski area mapping is recommended (Pröbstl 2001; Pröbstl et al. 2003). In a first step, the degree of hemeroby (i.e. the strength of human influence) is used for a first classification of the communities. Following comparable approaches, five main groups are differentiated:

- Natural communities, largely uninfluenced (ahemerobe)
- Replacement communities, largely natural (oligohemerobe) with limited influence
- Replacement communities, moderately natural (mesohemerobe) with moderate influence through human use
- Replacement communities, moderately unnatural (euhemerobe) with strong influence through human use
- Disturbed replacement and pioneer communities, unnatural or artificial (euhemerobe with strong disturbance/influence through site modification).

Aside from the degree of hemeroby, which is derived from the species list, further criteria considering the local characteristics need to be included in the valuation. Differences such as "alpine meadow" versus "alpine meadow with nutrient indicators around an alpine hut" should be represented, as should a *Cynosurus* meadow with an unusually high number of species.

The following parameters are used for further differentiation. Upgrading occurs in case of

- seedings with a tendency to develop towards higher quality communities,
- intensively used communities with occurrences of rarer species,

Table 4.9 Example of a description of ski piste vegetation in the ski area Schmittenhöhe, Zell am See

Geo montani-Nardetum (species-rich Nardus grasslands)	13
Site Widespread community of higher elevations; found on slopes of all expositions around the Schmittenhöhe, as well as between Sonnkogel, Schmiedhofalm and Sonnalm	
Description Nardus grasslands of different characteristics; grazed and ungrazed, often in contact with more alkaline pasture communities or near alpine rose heaths; dominated by matgrass; diverse, valuable, near-natural occurrences; one occurrence in a depression near the summit station of the Bruckberg lift with a lot of orange hawkweed	
Species array Gramineous species: *Nardus stricta*, *Festuca rubra* agg., *F. nigrescens*, *Luzula luzuloides* ssp. *rubella*, *Anthoxanthum alpinum*, *Luzula alpina*, *Juncus trifidus* ssp. *trifidus*, *Festuca rupicaprina*, *Trichophorum cespitosum*, *Luzula sieberi*, *Luzula multiflora*, *Luzula* cf. *sudetica*, *Calamagrostis villosa*, *Carex pilulifera*, *Deschampsia flexuosa*, *Phleum rhaeticum* Woody and herbaceous species: *Vaccinium myrtillus*, *Leontodon hispidus*, *L. helveticus*, *Homogyne alpina*, *Campanula barbata*, *Phyteuma betonicifolium*, *Crepis conycifolia*, *Hieracium auricula*, *Arnica montana*, *Potentilla aurea*, *Juniperus sibirica*, *Vaccinium uliginosum*, *Vaccinium vitis- idea*, *Hieracium alpinum*, *H. lachenalii*, *H. lactucella*, *H. stoloniflorum*, *H. aurantiacum*, *H, pilosella*, *H. laevicaule*, *H. intybaceum*, *Sedum alpestre*, *Phyteuma hemisphaericum*, *Veronica bellidioides*, *Melampyrum sylvaticum*, *Gentiana acaulis*, *Veratrum album*, *Gentiana punctata*, *Gnaphalium supinum*, *Cardamine resedifolia*, *Alchemilla alpina*, *Veronica officinalis*, *Prunella vulgaris*, *Achillea millefolium*, *Veronica chamaedrys*, *Ranunculus acris*, *Thymus pulegioides*, *Chaerophyllum villarsii*, *Lotus corniculatus*, *Platanthera chloranta*, *Calluna vulgaris*, *Blechnum spicant*, *Ajuga pyramidalis*, *Potentilla erecta*, *Thelypteris limbosperma*, *Pteridium aquilinum*, *Plantago lanceolata*, *Hypericum maculatum*, *Silene rupestris*, *Silene vulgaris*, *Dactylorhiza maculata*, *Hieracium aurantiacum* agg., *Rumex acetosella*, *Cirsium palustre*, *Trifolium pratense*, *Chrysanthemum leucanthemum*	
Significance/threat to the natural balance • Minimal damage through piste groomers, offers habitat to less competitive species • Greater damage on the slope edge near Schmiedhofalm (abrasion), a lot of Potentilla aurea at this site rare, protected species: *Trichophorum cespitosum*, *Dactylorhiza maculata*, *Platanthera chlorantha*, *Gentiana punctata*, *Gentiana acaulis*, *Hieracium stoloniflorum*, *Arnica montana*, *Vaccinium uliginosum*, *Alchemilla alpina* • Hab. 6230 • Value classification 8	

– mesohemerobe communities in a complex with higher quality communities.

Downgrading occurs in case of

– degradation due to different causes
– nutrient influx in natural communities
– damages due to construction measures
– large-scale damages due to grazing or impacted natural communities
– low degree of plant cover.

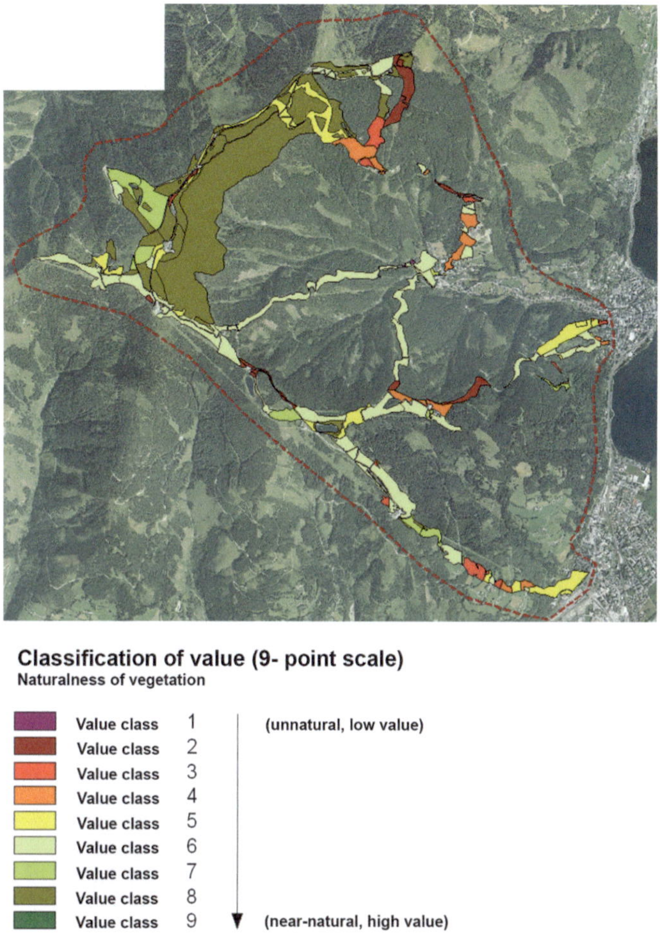

Fig. 4.10 Naturalness of vegetation communities in the ski area Schmittenhöhe in Zell am See/Austria (Pröbstl-Haider and Dorsch 2013)

Within the 9-point index, the five degrees of hemeroby can be assigned the following values, which can in turn be up or downgraded according to the above criteria depending on site-specific conditions (cf. Tables 4.10 and 4.11).

Size of the area or composite area covered by the community is not considered in the valuation.

Data on naturalness of the vegetation can be combined with other information in a geographical information system. This allows, for example, an analysis of the naturalness of plant communities on ski pistes according to altitude. This type of analysis has two aims: (a) the depiction of vegetation types by elevation level and (b) the depiction of the impact of management. Basic conditions for vegetation, such as temperature and duration of the vegetation period, change with increasing altitude.

Table 4.10 Classification of degree of hemeroby and scale of value (Pröbstl 2001)

Degree of hemeroby	Scale of value
Largely uninfluenced Plant communities	9, in case of impacts As low as 7
Largely natural plant communities Moderately natural plant communities	Between 6 and 8, base value 7 Between 4 and 7
Moderately unnatural	Between 3 and 6
Unnatural and artificial plant communities	Between 1 and 3

Table 4.11 Up and downgrading in valuation of the naturalness of different plant communities (Pröbstl 2001)

Up and downgrading in case of:	Extent
Degradation due to high nutrient influx or low species richness (depending on extent)	−1 or −2
Influx of nutrients Extensive grazing	−1 −1
Patchy vegetation cover	−1
Worthiness of protection of the habitat (e.g. protected by national law or listed in the Habitats Directive)	+1
Occurrence of red-list species	+1
Vegetation unit in complex with a higher quality community	+1

Furthermore, intensity of agricultural use usually decreases in higher altitudes, due to limited accessibility and possibility to use machines for management. What is more, this data can be used for interesting comparisons between the ski piste situation and other areas outside of the pistes. This allows e.g. an assessment of the natural regenerative capacity or the local degree of impact. Amongst other things, these comparisons have shown that the impacts at middle and lower elevations do not differ between pistes with and without snowmaking (Pröbstl 2006). Furthermore, the results indicate the success of recultivation measures or the effect of agricultural use of the area. The diverse possibilities for analysis and application as well as the high explanatory power for different aspects (soil, water balance, impacts and forms of use) justify the effort for these surveys.

4.4.4.2 Survey of Damages

Aside from the recording of plant communities there should also be a survey of areas where development of the vegetation is not possible or is impacted (Dietmann and Kohler 2005; Pröbstl 2001). Potential causes in ski areas are:

Impacts through ski edges, piste groomers, construction measures, grazing animals, hikers, mountain bikers, forest damages due to cutting, browsing damage from grazing or game animals as well as erosion damages.

Table 4.12 Proposed classification of damages along pistes in ski areas (Pröbstl 2001; Pröbstl et al. 2003)

Larger-scale damages in open spaces	Point-based and linear damages in open spaces in three classes	Damages in forest and mountain pine stands in three classes
Areas larger than 25 m^2	Areas up to 5 m^2 Areas up to 15 m^2 Areas up to 25 m^2	Areas up to 5 m^2 Areas up to 15 m^2 Areas up to 25 m^2
Areas without vegetation Areas with sparse vegetation (= vegetation cover between 15 and 50%) Other erosion events affecting larger areas (e.g. avalanche damages, pile-up, …)	Damages due to: – Summer tourism – Piste groomers – Construction and operation vehicles – Ski edges – Livestock trampling/turf damage – Water erosion – Avalanche blasting – Avalanche and snow gliding processes	Forest damages due to missing shelterbelt Windfall Breakages from wind and snow Damages through forestry use Damages through game animals Damages through grazing animals (browsing, root system, trunk)

In practice, it has proven essential to conduct such a survey in the winter months (especially towards the end of the season in March) and in summer. Extensive photo documentation is an important basis in this respect, because it can be a great help in classifying the damage causes. However, this method is usually only practical for point-based and linear damages. Aside from these forms of damage, there are frequently also larger-scale damages.

Table 4.12 shows a possible, well-proven classification scheme for damages in ski areas.

The survey of damages is also an important basis for the issues of soil, erosion and landscape damages (cf. Sect. 4.4.2).

Larger-Scale Damages

Areas with large extents of low vegetation cover of no more than 10–15% (proportion of the ground covered by plants in percent) at the time of the survey are defined as "areas without vegetation".

"Areas with sparse vegetation" are defined as areas with a vegetation cover between 15 and 50%. The time of the survey plays an important role in this respect. The assessment was not made in early spring, immediately after winter, but in summer (July/August). Thus, vegetation has had time to fill gaps, and the natural regenerative capacity of the site can be deduced. Another benefit of conducting damage surveys in summer is that impacts from summer use, e.g. damages from grazing animals or through summer tourism, will already have become effective. Thus, damages due to overlapping use can also be recorded (cf. Figs 4.11, 4.12 and 4.13).

Fig. 4.11 Damages related to grazing (*Photo* Ulrike Pröbstl-Haider)

Fig. 4.12 Damages related to mulch mowing (*Photo* Ulrike Pröbstl-Haider)

"Other erosions affecting larger areas" summarises all areas with large-scale erosion forms such as slope cracking, slumping or soil creep. These are mostly due to

Fig. 4.13 Damages related to the skiing operation (characteristic piste groomer marks) (*Photo* Ulrike Pröbstl-Haider)

natural causes (geologically unstable underlayers, overly steep slopes, etc.) or are the consequence of overlapping use or constructional modifications.

Ideally, the affected areas are mapped during the field surveys at a scale of 1:5000 (Pröbstl et al. 2003).

Point-Based Damages

In contrast to the larger-scale damages, which are mostly explained by a number of different causes (grading measures, ski edge abrasion, grazing, unstable sites), the linear and point-based damages can usually be attributed to single causes. Point-based damages are not only differentiated by cause but also by the extent of the damage.

The term "damages", in this respect, summarises a number of very different "disturbances" of the natural balance. Most damages are characterised by a strongly reduced vegetation cover (degree of ground cover less than 75%), often in combination with small surface scars. This goes for e.g. ski edge abrasion, grazing damages or impacts from hikers off the trails.

Forest damages include a very heterogeneous range of impacts. Damages due to sunburn at the forest edge or wind breakage can, for instance, be related to skiing and piste construction. Others, such as browsing damages by game or grazing animals, indicate potential additional impacts with an influence on the stability of the forest stands that are so important for the ski pistes and the natural balance. Contrary to the larger-scale damages, the small, point-based damages do not yet constitute long-

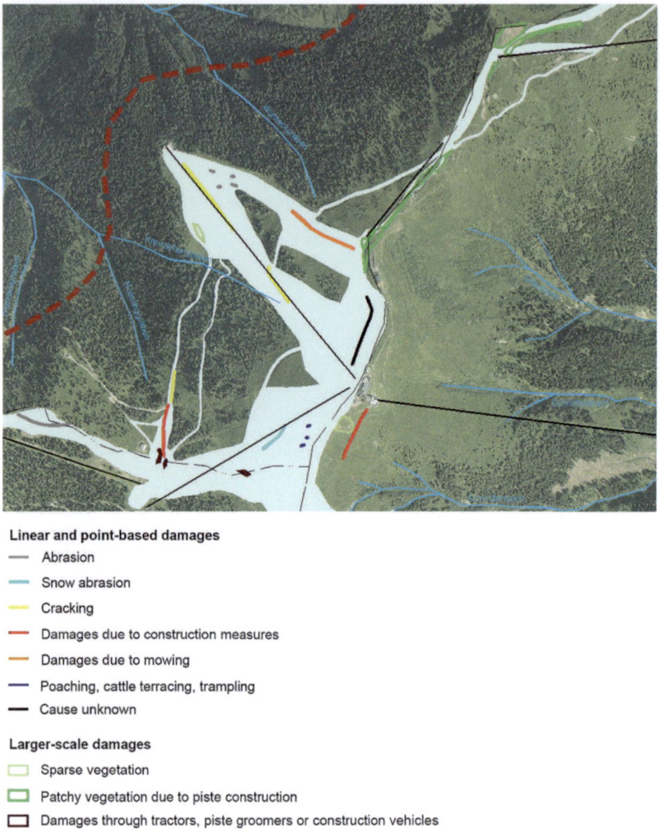

Fig. 4.14 Spatial mapping of linear, point-based and larger-scale vegetation damages; light blue = piste area (Pröbstl-Haider and Dorsch 2013)

term disturbances. However, surveys in Bavarian ski areas (Pröbstl 2001) showed that a differentiated recording of point-based damages, in particular, offers a special indicator function, allowing the elaboration of targeted restoration proposals and also a better understanding of the underlying causes of larger-scale damages.

The survey is a snapshot of point-based damages that can usually be eliminated in a few years through targeted restoration or protection measures.

The analysis of damages in the ski area should illustrate the most common types of damages and extent of affected area. This data can then be compared with the results from the ski pistes. The size of the affected area is ascertained via databases containing the extent of damages according to the classification (point-based damages) or the local delimitation (larger-scale damages).

Measures can then be planned based on the survey of damages (cf. Fig. 4.14). Their effectiveness can also be documented and reviewed in an environmental management system, for example by comparing vegetation surveys from previous years.

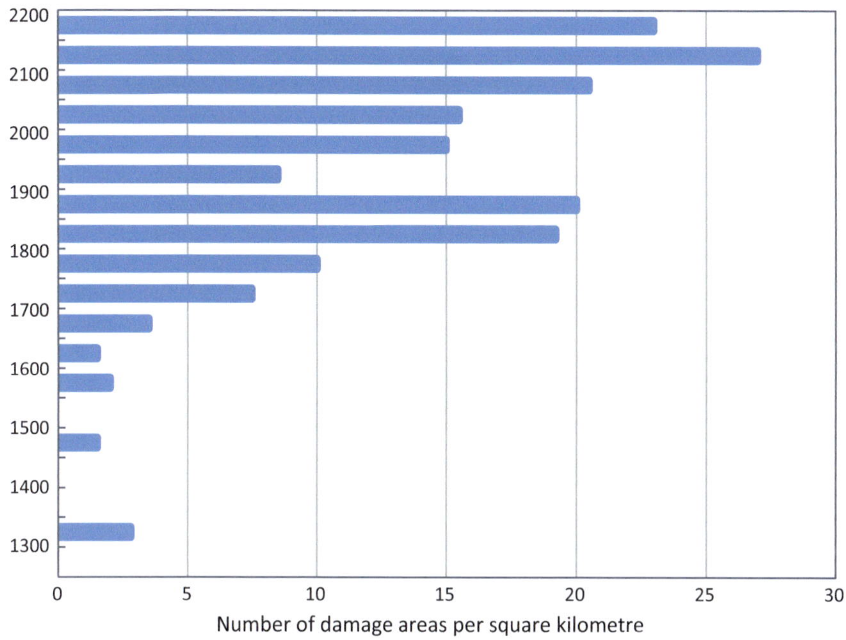

Fig. 4.15 Distribution of damage areas according to elevation in the survey area Adelboden (Pröbstl et al. 2003)

The analysis according to elevation (cf. Fig. 4.15) shows that damage areas increase with higher elevations. This is an important signal calling for action, since the natural regenerative capacity decreases with higher elevations. Active measures are required in these cases.

Regarding ecological quality, improvements to the ski pistes and bordering forests can be achieved through the following measures:

- Re-greening of surfaces, construction roads or paths that are no longer required,
- Control and eradication of invasive species by mowing and subsequent removal of cuttings,
- Review of the use of fertilisers and reduction as far as possible,
- Implementation of extensive management through mowing or grazing and refraining from mulch mowing,
- (Temporary) protection of valuable habitat areas through fencing off,
- Visitor guidance in areas with sensitive vegetation and risk of erosion,
- Limiting driving across sensitive piste sections or already damaged areas,
- Re-greening of damaged areas through seeding or turf transplantation,
- Preserving habitat trees (fulfilling an important habitat function, e.g. old trees with hollows) under consideration of the legal obligation to maintain safety,
- Improving information on causes and consequences of forest fires for the public (e.g. through signs, on the website or orally) and also in personnel training,

Fig. 4.16 Native birds should be included in the review. In mountain areas, grouse such as the Western capercaillie are of special relevance (*Photo* Frank Armbruster)

- Improving cooperation with local farmers to ensure use of the mowed material,
- Review of the inventory of trees with a view to ecological inclusion and improvement of parking areas and buildings.

4.4.5 Fauna

Regarding potential impacts of winter sports in ski areas, those species must be considered, above all, that can suffer potential serious impacts and that are protected by species conservation laws (e.g. according to European or national law). This often applies to bird species, particularly those that are active in winter (Zeitler 2001) (cf. Fig. 4.16). Special consideration must also be given to amphibians near bodies of water or wetland terrain. The habitats of species listed in the Habitats and Birds Directives need to be examined during the environmental management. The Appendix lists species from Annex IV of the Habitats Directive requiring protection in the Alpine region.

Information can be sourced from local experts, authorities (conservation and forestry authorities), an evaluation of official documents (e.g. species and habitat protection programmes) and personal observations or specific surveys (cf. Table 4.13). Since separate surveys involve a greater effort, they should be decided on depending on the significance of the potential impacts and the relevance of the intended land use for a respective species. Often the statements of local experts and an analysis of existing surveys and habitat analyses are sufficient.

Depending on the specific situation, the following bird groups should be examined in particular:

Table 4.13 Data assessment for the subject of protection fauna (adapted from Pröbstl et al. 2003)

Fauna			
Data sources and possible contact persons	Basic information	Priority	Reason and comments
Conservation laws of respective countries or provinces, other laws Designation of protected areas Range maps Protected areas and descriptions (incl. habitat conservation forest) Red-list species of the area Landscape planning projects and other larger-scale planning projects Existing surveys Consultation of local experts in the fields of forestry, hunting and conservation	Ranges of (protected) wild animal species	A	Basis for ascertaining potential sources of conflict and improvement measures
	Game reserves	A	
	Ranges of protected bird species (leks)	A	
	Habitat structures (forest/open spaces)	B	Improvement of habitat potential through a change of use during the course of improvement measures
	Avalanche blasting cableway	B	Possible impact on game animals

- Species of forest edges, if pistes and facilities permeate the forest (winter residents)
- Diurnal cavity breeders (woodpeckers in particular), if pistes and facilities permeate or are close to forest
- Grouse in forest or above the treeline
- Owls and raptors, if pistes and facilities permeate the forest and snowmaking is conducted at night.

In order to minimise the effort of field surveys, potential habitats of bird species occurring in the area should be defined before the survey rounds. In most cases, local experts (e.g. foresters, ornithologists, employees of the business) can provide sufficiently detailed information to narrow down the area for the field surveys. In some cases it might be possible to draw on data from EIA procedures or other approval procedures.

Usually 2–3 survey rounds are sufficient to describe the important species and their home ranges in the quality needed to determine further targets and measures (Renat 2000).

As Table 4.13 shows, the available details on animals can be used as a basis for deriving concrete measures. In the case of grouse (Western Capercaillie, Black Grouse, Rock Ptarmigan, Hazel Grouse) there is occasional conflict with the skiing facilities (e.g. the cables of tow lifts). Improvements can be planned in these cases. Regarding the risk of bird collision, cable thickness is a relevant factor. With a cable thickness of 5–15 mm, collision prevention is not possible, or only possible for small to medium-sized bird species. With a cable thickness of 15–35 mm, collision

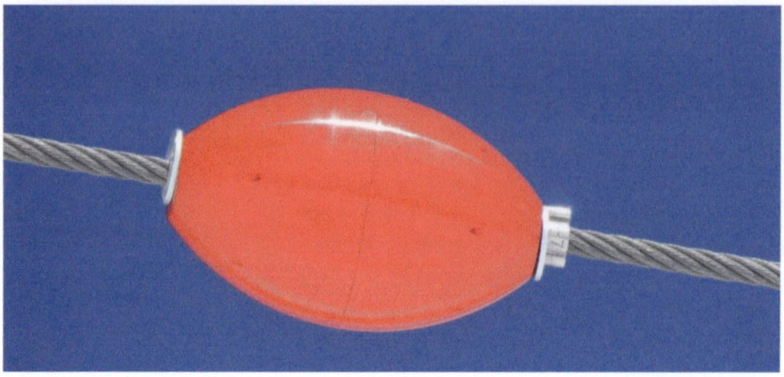

Fig. 4.17 Protection of wild birds through improved visibility of cables (*Photo* Sandrine Berthillot)

prevention by means of colouring or thickening the cable is possible. With a cable thickness of more than 35 mm, the risk of collision is lower and is limited to twilight or low visibility due to fog.

Location and vegetation as well as lay of the terrain and the lift route also play a role. Routes near foraging sites, preferred resting areas or escape routes pose a particular danger (Zeitler 2014).

Recommendations by various cableway operators show that measures to increase cable thickness and improve visibility through colouring are effective. A common recommendation is to coat the cable in a flexible PVC electrical installation tube (32 mm, with slits, 50 m per segment). Ice coating was not observed, due to the vibrations (Manhart 2014). Increasing visibility by attaching a blue cable is further recommended, because this colour is perceived well by birds, even in forest.

Use of protective markings available for overhead power lines, e.g. pairs of black and white spirals, has not yet been tested. Spacing would need to be short in order to be effective (10 m) because birds cannot "interpolate" and fill in the line (Zeitler 2014; Kalz et al. 2015).

Experience from French ski areas confirms the need to make dangerous elements of cableways more visible in order to protect birds. Here, a protective effect and reduction of killed birds was observed after use of red plastic balls along the cables (cf. Fig. 4.17). It became apparent in this case that Black Grouse and Rock Ptarmigan are disproportionately affected species that benefit in particular from such measures (Conseil général de la Savoie 2004).

Furthermore, during management measures in France, all fences were also critically examined. A potential for danger was apparent here as well. In this case, the risk of collision could be reduced significantly by applying metal or plastic disks. These little discs, applied in regular intervals, again increase visibility for the animals. All of these measures are accompanied by information for the general public.

In his analysis of bird collisions at ski stations, Zeitler (2014) came to the conclusion that measures to greatly improve visibility and coloured blinds (20 cm wide horizontally and 10 cm spacing) were well suited to preventing bird collisions on window panes. The Swiss Ornithological Institute found similar results for efficient structuring and design (see Schmid et al. 2012).

It is important, in this respect, to consider the special characteristics of ski areas. During operating hours, when the facilities are heavily frequented, the risk of bird collision is very low. The risk is by far higher in the early hours of the morning, in the evenings and at night, and can be exacerbated by adverse weather conditions such as fog.

But experience from Germany, Austria, Switzerland and France also shows that an area-specific assessment of risk should be pursued. From the experts' point of view, the following aspects can make a great difference from one area to the next (Zeitler 2014):

- the topographic conditions around the facilities as well as the surrounding vegetation and habitat quality, the elevation and bird species occurring in the area along the entire lift routes, and the locations of the buildings,
- the distribution of preferred habitat structures in the area,
- the effects of sun and shadow on the cables and buildings during the course of the day and year (reflections and dark shadows) as well as the sun's angle of incidence depending on the time of year,
- exposure of buildings and choice of materials (especially glass),
- the foraging and resting areas of the respective bird species,
- the daily centres of activity of the occurring bird species,
- the resting areas and escape routes in case of disturbance through predators or humans.

These factors can add up to very different risks of collision.

Aside from the risk of collision along the cables, in the case of cableway operations, the use of glass in the buildings also plays an important role. There are numerous design possibilities in this case (Zeitler 2014) permitting a bird-friendly use of glass. These include the inclination of the panes to reduce reflections, as well as the use of colours and structures on the glass. The use of blinds can have a further preventive effect, because they can reduce the risk of collision even in bad visibility such as twilight, fog, etc.

As mentioned above, other animal species aside from the bird species should also be included in the review. This goes especially for species of Annex IV of the Habitats Directive. In practice this mostly pertains to the different native bat species that can be affected by construction or renovation work on buildings or impacts on the inventory of trees. With appropriate planning and design, the habitat structure for bats can even be improved.

The species demanding consideration according to Annex IV of the Habitats Directive in the ski area Schmittenhöhe included numerous protected amphibians and reptiles. Special consideration was given to their habitats during management. Details were gained from an assessment completed for an EIA procedure.

Further possible measures for the protection of wild animals can include the following:

- Adding habitat areas near bodies of water for snowmaking;
- Consideration of wild animals during planning of construction measures, summer events or major sports events;
- Attention to permeability of terrain during planning and maintenance of guards and safety fences;
- Improving habitat structures in less disturbed areas.

4.4.6 Water

When assessing the hydrological situation, the location of the area in the catchment areas of different trough systems, mountain streams and recipients should be investigated. Furthermore, water protection areas, spring catchments, local water rights and agreements of use should be observed. The issue of water extraction for snowmaking must also be addressed in this context (cf. Table 4.14). This includes the type and quantity of water extraction, the delimitation of areas to which artificial snow is applied and information regarding the average times during which snow is applied. Points of water discharge from the piste must be examined carefully, especially if the piste was graded and snow is applied regularly. They can constitute starting-points for erosion processes.

Since climate change can have an impact on the local water balance (Böhm et al. 2008; Blaschke et al. 2011; ÖWAV 2014), monitoring of bodies of water in ski areas is of particular significance. Precipitation in winter has increased slightly since the 1970s north of the Alpine divide, but has fallen considerably south of the Alpine divide (ÖWAV 2014). Thus, winter floods must be reckoned with more frequently and altered drainage properties around bodies of water are to be expected. Regarding filling of reservoir lakes, more frequent low water situations, particularly in the southern Alps, must be considered.

In ski areas, especially where snowmaking systems are used, a differentiated view of matters concerning the subject of protection water is necessary. Regulations to this effect are also found in the permit documents. Constant monitoring provides an important basis, which can be evaluated during the auditing process. Important issues are, whether and to what extent:

- water is available in sufficient quantity and quality,
- it is ensured that neighbouring bodies of water, catchments and usage rights or spring areas remain unaffected,
- installations in the body of water allow potential floods to be discharged safely,
- habitat connectivity and quality for aquatic organisms is considered and
- the permissible extraction quantities are observed.

It is recommended, in this respect, that the business maintains its own local series of measurements. The following parameters are of particular importance: spring yield, discharge (lowest, average and maximum) as well as quality parameters.

Table 4.14 Data assessment for the subject of protection water (adapted from Pröbstl et al. 2003)

Water			
Data sources and possible contact persons	Basic information	Priority	Reason and comments
Mapping of bodies of water Mapping of water protection areas Water management authorities Cableway operators Permits for water extraction and pipeline courses Cadastre of springs and catchments Index of municipality drinking water catchments	Bodies of flowing water	A	Frequent interdependence with water discharge and hydrological balance of the piste, significance for the formation of erosion processes and mudflows, consideration of retention areas
	Bodies of standing water	B	Potential valuable habitat, especially if spring fens are involved
	Water protection areas (drinking water catchments)	C	Valuable potential Resource requiring protection
	Catchment areas	A	Basis for ascertaining the consequences of piste preparation and additional water for snowmaking
	Water for snowmaking	A	Key parameter for the skiing operation and potential impact on the natural balance

When planning measuring facilities, their location should be chosen in relation to the locations of existing authority measurement facilities. Thus, own measurements can be integrated with public data and further reaching interpretations are possible.

Starting points for environmental management and improvements in this respect are:

- implementing efficient snowmaking systems and a differentiated management system;
- constructing reservoir lakes to relieve natural or near-natural bodies of water;
- using GPS-based measurement of snow levels to help save use of artificial snow;
- use of computer-controlled snowmaking systems to optimise snow quality and water demand;
- installation of water-saving infrastructure, e.g. in toilet facilities and restaurants;
- avoiding landscaped areas that require watering;
- improvement and monitoring of waste water facilities and systems;
- training personnel in the field of snow management and
- training personnel on matters of water consumption and means of saving water in buildings.

Table 4.15 Data assessment for the subject of protection humans (immissions)

Humans (immissions)			
Data sources and possible contact persons	Basic information	Priority	Reason and comments
Roadworks authorities/administration Immission control in regional administration Mapping of snowmaking systems Traffic concept and delivery logistics Mapping of illuminated areas	Pollution along adjacent roads (initial level of pollution)	B	Important data for cumulative effects, e.g. snowmaking and indirect effects of traffic
	Location of snowmaking systems and deduction of an area affected by noise	A	Protection of local inhabitants and tourism businesses
	Business-induced impacts due to traffic noise	B	Important basis for induced impacts that also require measures
	Light pollution due to illumination at night	C	Potential disturbance of residential areas

4.4.7 Immissions (Noise/Light)

The regular skiing operation, organised tournaments and snowmaking may cause conflicts due to noise immissions. Increased volumes of traffic and further areas affected by noise are to be expected due to piste preparation, delivery and collection e.g. of goods and fuels, as well as through the personal transport of the guests themselves.

These aspects are generally regulated in the permit for the installation, but they could become relevant if installations or systems need to be exchanged or replaced.

The noise immissions caused by snowmakers, which are particularly relevant, can be visualised in the basic maps, thus allowing affected or critical areas to be identified early on.

The local inhabitants, but also tourist operations in the area, enjoy legal protection.

Aside from snowmaking, operation of facilities for nighttime skiing can also cause stress and conflict due to extensive illumination. These particular adverse effects can differ markedly from case to case, depending on forest cover, number of trees, topography and building development. Therefore, no general recommendations are made in this case. Possible data sources for assessing the noise and light immissions in a ski are listed in Table 4.15.

4.4.8 *Waste*

Different cycles and the development of waste have to be observed within the business. This includes

- waste created by the guests, such as food scraps, packaging, etc., but also broken ski glasses and other sports equipment or lost articles of clothing,
- waste created through operation of the installations, especially the fleet of vehicles,
- waste created through operation of guest facilities and associated infrastructure, such as ski rentals or sports stores.

The respective types of waste need to be assessed separately for each of the different sectors. More information on assessing the issue of waste in a ski area is given in Tables 4.16 and 4.17 shows an according practical example.

Starting points for further improvements in the business are:

- conversion to a high proportion of recyclable products;
- conversion to large and refillable units,
- switching to suppliers with a take-back service
- training of personnel;
- improving information for guests and conducting an examination of locations and facilities (in some cases reducing the number of rubbish bins actually increases the willingness of recreation seekers to take their rubbish back with them);
- largely abstaining from use of one-way materials, even in gastronomy, and
- an examination and improvement of waste separation facilities and of subsequent recycling options.

Table 4.16 Data assessment for the issue of waste

Waste			
Data sources and possible contact persons	Basic information	Priority	Reason and comments
Information provided by disposal services E.g. extent of recycling take-back Information of individual departments on collected waste Information of individual sectors of the business (e.g. shop, restaurants)	Compilation according to type and quantity of waste as well as disposal requirements and recyclability	A	Information on quantity is required in order to achieve a reduction Type of waste allows assessment of recyclability Type of waste determines disposal options
	Compilation according to sectors Compilation according to self-contained sectors of the business	B	Allows direct allocation of measures Allows inclusion of suppliers and reduction of waste through chosen type of packaging/delivery

Table 4.17 Example of waste data and disposal (Schmittenhöhebahn AG 2014)

Period: 01 Dec 2009–31 Nov 2010 (1 year)			
Company	Type of waste	Source	Quantity
Gassner disposal	Bulky waste	General	0.74 t
Gassner disposal	Construction and demolition wood	General	2.42 t
Zemka	Garden and green waste	General, mountain huts	51.75 m^3
Zemka	Mixed commercial waste	General	20.38 t
Zemka	Scrap wood	General	0.84 t
Zemka	Scrap iron	General	0.25 t
Zemka	Waste glass	General	0.12 t
Zemka	Construction rubble	General	6.96 t
AVE	Organic kitchen waste	Gastronomy	19,440 l
F-S GmbH	Construction water	General	25.0 m^3
F-S GmbH	Waste oil	Restaurants	2500 l
F-S GmbH	Oily waste	General	556 kg
F-S GmbH	Oil and grease separators	Restaurants	75 m^3
F-S GmbH	Antifreeze	General	39.0 kg
F-S GmbH	Cleaner solvent	General	45.5 kg
Municipality Zell am See	Organic waste bags	Areit cable car, gastro, admin	240 l
Municipality Zell am See	Organic waste-paper	Areit cable car, Schmittenhöhe	4180 l
Municipality Zell am See	Organic waste bin	Office buildings	2080 l
Municipality Zell am See	Non-recyclable waste bin	General	212,940 l

The compilation allows targeted measures to be developed in order to achieve a lasting reduction of the quantities of waste produced.

4.4.9 *Use of Resources*

4.4.9.1 Operating Materials, Material Usage

The operating materials used in ski areas largely include fuel, oils, fats (e.g. lubricants for maintenance), cleaning materials, varnish, paints, hardeners, fertilisers (e.g. organic fertilisers such as Biosol and stable manure), batteries, cooling and antifreeze agents.

The fleet of vehicles and piste groomers are the main fuel users, followed by other vehicles (e.g. snowmobiles and construction vehicles). In order to ensure efficient material usage, the conditions of usage should be reviewed and optimised where necessary. Especially when it comes to fuel consumption, reducing and optimising the number of trips necessary can lead to great savings. Efficient material usage should be documented in the environmental statement with appropriate key figures. One such key figure, for example, could be fuel consumption or material usage in kg or t per hour of operation.

Surveys for the audit of the Skilifte Lech (Ing. Bildstein Ges.m.b.H.) showed that 70% of diesel consumption was due to piste preparation. This knowledge allows measures for piste maintenance and management to be deduced, such as later preparation after fresh snow or the gradual expansion of prepared surfaces while retaining areas of deep snow.

Converting to biodiesel would mean further improvement of the CO_2-balance, although the possibly limited winter suitability of biofuels would need to be considered. It is certainly a feasible option for maintenance during the summer. Depending on the origin, the CO_2-savings compared to regular diesel are 38% in the case of rapeseed, 51% in the case of sunflower seed, and 83% in the case of biodiesel from plant and animal waste (the figures are based on directive 2009/28/EC of the European Parliament and of the Council of 23 April 2009).

Reviewing piste preparation effort by evaluating the GPS-based records of piste groomers (e.g. during snow measurement) can indicate ways to increase efficiency or achieve best practice. In this respect, comparisons between different piste sections and also between different businesses using this technology can also provide starting points for improvements within the business and can help learn best practice.

4.4.9.2 Energy

Ski areas in European Union member states face a concrete set of challenges thanks to the European Energy Efficiency Directive (Directive 2012/27/EU). But cableway operators will also have to consider further directives in the future, such as the Directive on the energy performance of buildings (2010/31/EU) and the Directive on the promotion of the use of energy from renewable sources (2009/28/EG). The Energy Efficiency Directive (2012/27/EU) states that larger businesses and organisations with public involvement must conduct energy audits. The significance of environmental management systems (e.g. EMAS, ISO 14001) is increased, since their implementation typically includes such an energy audit.

As with the previous issues, the basic assessment for the issue of energy must be based on the question of whether the data assessed is suited to lead to concrete measures. The appraisal should provide the basis for the relevant sectors:

1. How much energy is required and how can the current energy demand be reduced?
2. Which sectors are particularly energy intensive?
3. Where is energy currently sourced and could it be sourced in a more environmentally friendly way (e.g. certified green energy)—especially during peak times?
4. Is energy produced by the business itself or would this be a feasible option for the future? How high is the current proportion and could it be expanded?

5. To what extent can cooperation/participation and marketing lead to improvements?

Generally, the following sectors of relevance for a business' total energy consumption need to be differentiated:

- Piste preparation
- Snowmaking
- Lift systems
- Indoor heating
- Gastronomy
- Infrastructure and technical facilities.

The following example of the Skilifte Lech (Ing. Bildstein Ges.m.b.H.) illustrates which measures concerning the issue of energy are possible:

- Participation in biomass heating plants
- Heating by geothermal energy
- Installing a photovoltaics system and integrating a solar system for water heating
- Reducing energy demand through waste heat recovery
- Using high-grade insulation materials in buildings.

The great majority of energy in Lech (approx. 52%) is used for snowmaking, approx. a third is used for the cableway systems and 9% for gastronomy. Since energy demand for snowmaking is relatively high, measures should be implemented in this area in particular. These can include snowmaking only when it is absolutely necessary, and only producing the quantities that are absolutely necessary based on experience. Differentiated snow management is a further help (e.g. using GPS-based solutions for measuring snow depth).

The efforts of the Skilifte Lech in 2014 to enforce renewable energy led to a photovoltaics system installed on the roof of the office building, producing an average of 8.850 kWh of electricity per year. The energy expenditure needed to manufacture the system is thereby amortised after approx. two years.

Despite the extensive complex of buildings and offices, requiring about 800,000 kWh in total, the audit for the Skilifte Lech also showed that there is barely potential for improvement in terms of CO_2-equivalents, because the oil heating has already been replaced by community woodchip heating. Heat pumps (geothermal energy) and solar water heating are also used.

Figure 4.18 shows that, even for other cableway businesses, electricity consumption offers the most potential to achieve a reduction of cost and to implement environmental management measures. Here again, snowmaking and piste preparation are issues with a high savings potential.

Measurements and extrapolations of snowmaking based on a comparison of the winter seasons 2008/2009 and 2011/2012 in the German ski area Fellhorn-Kanzelwand showed that considerable savings could be achieved thanks to snow depth measurement (Schuster 2016). Due to this increase in efficiency, snow depth

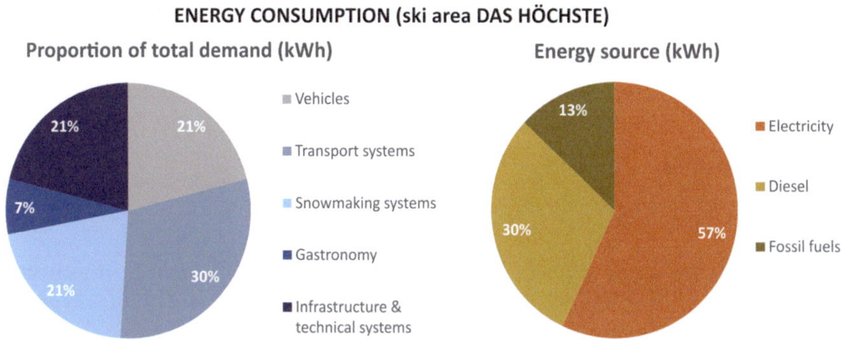

Fig. 4.18 Energy consumption in the ski area "Das Höchste" (Walmendingerhorn, Ifen, Kanzelwand, Fellhorn- und Nebelhornbahn) (own diagram based on Oberstdorf/Kleinwalsertal Bergbahnen 2017)

measurement systems are expected to become standard in modern ski areas. The test series have shown that the initial investment is amortised by the possible savings in one to two seasons (depending on the size of the ski area).

As mentioned above, further possible savings can be achieved by:

- use of renewable energy, e.g. hydropower, geothermal energy, photovoltaics and solar heat;
- installation of control systems for lighting, such as timers and motion sensors as well as energy-efficient lights;
- regular examination of the buildings' heating and cooling systems and improvement of the control systems wherever necessary;
- training of personnel in the field of energy-efficient building management;
- improvement of insulation in heated buildings;
- use of snowmaking and water reservoir systems to produce energy and
- optimisation and renewal of the snowmakers and control systems.

In developing measures, it can be of help to convert energy-related services into key energy figures, thus making them comparable. Energy-performance-indicators based on "skier days" (kWh/skier day) or per lift system (kWh/transported metre of elevation) can be of great assistance in this respect.

4.4.9.3 Indirect Environmental Impacts

As has already been mentioned, indirect environmental impacts also need to be addressed. Indirect environmental aspects cannot be fully influenced by the ski area and are often related to other service providers, supply firms as well as personnel. Furthermore, ski areas only have limited influence over the behaviour of their winter sports guests. Nevertheless, it is also up to the environmental audit to try to influence the environmental performance of contractors, subcontractors and suppliers, or to

Table 4.18 Means of arrival in the ski area Schmittenhöhebahn (Zell am See) (Ecosign 2005)

Means of arrival within the resort	Proportion in %
On foot in walking distance	Approx. 34%
By car	Approx. 42%
With the local ski bus	Approx. 24%

work towards an improvement. This includes transportations, means of delivery, packaging, etc.

Other indirect impacts are influenced by capital investment and loans as well as administrative and planning decisions. Regarding CO_2-balance, means of travel to and from the ski area, above all, belongs to the significant indirect environmental impacts that need to be evaluated regularly (cf. Table 4.18).

Even if the business has no role in deciding means of travel, it can try to exert an influence on the decision, e.g. using incentives. These can include cooperations with local public transport services, railways or bus companies. An example from Zell am See shows related efforts to reduce the proportion of guests arriving by car. The offer does not only include parking spaces, but also arrival options by train and a ski bus for the ski regions Zell am See and Kaprun. Regarding guests staying in the resort, means of arrival is divided as follows:

The following measures are regularly seen as starting points for improvement:

- developing incentive systems for employees to make their commute to work more environmentally friendly, e.g. through use of public transport or carpooling. The business can incentivise these measures with bonuses, covering of transport costs, business-own shuttles, etc.
- developing incentive systems for guests, e.g. through cooperation with local public transport services.
- cooperation with accommodation businesses regarding pricing (dates of arrival) to reduce traffic peaks at the weekends.
- joint development of car-less offers with tourism associations and travel operators.
- cooperation with the municipality in developing a traffic concept aiming at a car-free ski area.
- cooperation with travel agencies to encourage tourists to arrive by train.

4.4.10 Data Assessment by Form of Use During First Environmental Review

A detailed assessment of the forms of use is seen as an important part of any ski area mapping. The recommendations of comparable studies differ in this respect (Leicht et al. 1993; Dietmann and Kohler 2005). The most important forms of use in ski areas include:

- winter sports,
- summer tourism,
- agriculture and alpine farming, as well as
- forestry and hunting.

Detailed mapping of the forms of use is important because it is often not the individual impact that is problematic, but the cumulative impacts of different forms of use. A further "special form of use", if not in the classic sense, is nature conservation. Protected areas, valuable and mapped habitats must therefore be surveyed and illustrated separately. This aspect has gained in significance in European Union member states in connection with the Habitats Directive and the Birds Directive.

Forms of use that are interdependent with winter usage must be examined and surveyed in detail. These include grazing of ski pistes or hiking along trails in summer, for example, but also usages belonging to or emanating from the business itself, such as selling sports equipment, operating restaurants or organising sports events. These areas cannot be described here in a standardised way. A compilation of aspects that need to be examined regularly is shown in Table 4.19. The classification of priority in classes A–C is explained in Sect. 4.5.2; A stands for high priority and C for lower priority.

Checklist for environmental impacts

□ Does the assessment and evaluation of environmental impacts take all affected subjects of protection and forms of resource use into account?

□ Are the following environmental aspects adequately considered?

- Condition of vegetation
- Protected animal species
- Erosion, soil creep and other signs of damage
- Protection of bodies of water and drinking water
- Microclimate and adaptation to climate change
- Noise
- Lighting, optical effects
- Waste
- Use of resources (water, energy, basic materials).

□ Are the central environmental impacts adequately documented in text and maps?

□ Is there an adequate and regular stocktaking of used resources?

□ Are indirect environmental impacts also assessed and taken into account?

Table 4.19 Data assessment for infrastructure and form of use (adapted from Pröbstl et al. 2003)

Infrastructure and form of use			
Data sources and possible contact persons	Basic information	Priority	Reason and comments
Business/Ski area operator Literature Interviews and application documents National tourist office Municipality Farmers' association Grazing cooperative Cableway operators Forestry office Sports office Conservation authority Internet Societies Hunting association	Demarcation of ski pistes	A	Basis for assessing impacts
	Lift systems (by year of construction, capacity, type, average occupancy, significance summer-winter)	A	
	Buildings (Mountain huts, guesthouses)	A	
	Roads, paths and trails	A	
	Areas with artificial snow, locations of the snowmakers	A, Locations B	Basis for assessing impacts, potential significance for structural modifications
	Areas relevant for sports tournaments	A	Particular intensity of use
	Special winter sports areas (funpark, toboggan runs, ski kindergartens)	A	Basis for potential additional impacts
	Protected areas, officially mapped habitats	A	Measures for protection and promotion
	Frequency of use of paths, pistes and facilities	A	Basis for ascertaining impact
	Infrastructure for summer use (take-off points for hang-gliders, biking routes, playgrounds)	B	Potential further impacts
	Parking areas	A	Basic data/map
	Agricultural use (buildings, fences, grazing, mowing, livestock density per ha, animal species, fertiliser)	A, Details B, C	Sections/areas with potential further impacts
	Forestry use	A, C	Forest units with a protection function have a high priority, other forests a low priority
	Hunting use	B	Conflicts of interest with recreation use possible
	Transmission routes (e.g. waste water, electricity)	B	Potential disturbance of the subsurface through waste water, potential impact on animals of power lines

4.5 Evaluation and Identification of Need for Action

4.5.1 Data Preparation and Data Management

4.5.1.1 Spatial Data

Since data with a spatial reference is used both in establishing the environmental management system and in ski area management, combining the data in a geographical information system (GIS) is recommended, especially for larger ski areas. This makes different content readily available, allows data analyses, combinations and visualisations as well as easy updating of the data.

Through integration of further data, such as newly planned projects, such a system becomes the ideal basis for middle and long-term development of the ski area, as well as for queries and derivation of planning documents.

Creating a base map with all fundamental information for the area is recommended during the course of the first environmental review.

The map, which can be supplemented by a digital orthophoto, should contain the following information (Pröbstl et al. 2003):

- Lift systems (cable cars, chairlifts, etc.)
- Piste demarcation (in a second step, if necessary)
- Buildings (infrastructure of the ski area, residential buildings, huts, etc.)
- Demarcation of forest and open spaces, mountain pine stands and rocky areas (orthophoto)
- Network of paths (classification according to the base map, parking areas)
- Contour lines (generated from the digital altitude model, if necessary)
- Bodies of flowing water and standing water, springs
- High-voltage power lines, prominent points for orientation in terrain
- Administrative borders
- Demarcation of protected areas, if applicable.

A lot of general data can be obtained from land surveying institutions and can be added to the base map.

General base data (aerial images and orthophotos) can usually be obtained from the national land surveying institutions of the respective countries.[3] A lot of technical information, e.g. regarding slope stability or protected habitats, is already available in digital form.

[3]In Austria, for example, such data can be obtained from the Bundesamt für Eich- und Vermessungswesen (BEV).

4.5.1.2 Content Data

A lot of the content data is based on quantifiable results. These can pertain to the operating materials used, the waste generated or to water and energy consumption. Since the data is mostly present in table form, it is recommended to visualise the data e.g. in bar charts and to compare long-term trends. This also makes it easier to formulate threshold values and environmental objectives to pursue. Examples are illustrated for the issue of energy.

4.5.1.3 Assessment of Legal Bases

As mentioned in Sect. 4.3, a ski area has to pay attention to a number of very different legal matters in order to act in conformity with the law. ISO 14001 prescribes an assessment and the observation of legal requirements. It is therefore necessary to assess all legal obligations regularly, to evaluate them and to prove that they are abided by. Ideally, a ski area should compile all applicable laws and obligations in a so-called legal database.

In Austria, the relevant legal bases can be found in the legal information service of the federal chancellary (Rechtsinformationsdienst des Bundeskanzleramtes—RIS, https://www.ris.bka.gv.at/). The Austrian chamber of commerce (Wirtschaftskammer) also offers their so-called calendar for business owners, which informs about recent changes in environmental law (see: https://www.wko.at/service/umwelt-energie/unternehmerkalender.pdf). Sector-specific legal information is also provided by the respective trade associations of the chamber of commerce. The trade association of cableway systems also informs its members regularly about significant new developments in environmental law (e.g. EIA legislation, species conservation law, water law). International organisations such as OITAF[4] also provide supportive technical resources. On top of this, further legal matters regarding personnel protection and safety as well as requirements under employment law need to be considered.

4.5.2 *Evaluation of Environmental Aspects*

It is important to evaluate the direct environmental impacts specific to the ski area situation, and to provide management with reference points regarding urgency and prioritisation. A so-called environmental database is a helpful tool in this respect, representing all aspects in table form and summarising potential risks regarding different natural subjects of protection and use of resources. Aside from potential risks, the current state is also assessed.

[4]The International Organisation for Transportation by Rope, established in Milan/Italy in 1959 (**o**rganizzazione **i**nternazionale **t**rasporti **a** **f**une).

Environmental impact / Aspect	Soil		Water		Flora		Fauna		Air		Noise		Waste		Use of resources: Packaging		Water		Energy		Operating materials		Need for action
	1	2	1	2	1	2	1	2	1	2	1	2	1	2	1	2	1	2	1	2	1	2	
Snowmaking	C	B	B	A	B	B	B	B	C	C	A	A	C	C	C	C	A	A	A	A	C	B	III
Winter piste maintenance	B	B	B	B	B	A	C	B	C	C	B	B	C	C	C	C	C	C	B	B	B	A	III
Skiing off-piste	C	C	C	C	B	A	C	B	C	C	C	B	B	B	C	C	C	C	C	C	C	C	III
Summer piste maintenance	B	A	A	A	A	A	B	B	C	C	C	C	C	B	C	C	C	C	C	C	C	C	I
Cableway operation	C	C	C	C	C	C	C	C	C	C	C	C	B	C	C	C	C	C	A	A	B	B	III
Gastronomy	C	C	B	B	C	C	C	C	C	B	C	C	B	B	B	B	B	B	C	C	C	B	III
Shop/sales	C	C	C	C	C	C	C	C	C	C	C	C	C	B	B	B	C	C	C	C	C	C	IV
Skirental	C	C	C	C	C	C	C	C	C	C	C	C	C	C	C	C	C	C	C	C	C	C	IV
Administration	C	C	C	C	C	C	C	C	C	C	C	C	C	C	C	C	C	C	C	C	C	C	IV
Waste water treatment	C	B	A	A	B	A	C	B	C	B	C	C	C	C	C	C	C	B	C	C	C	C	III
Drinking water supply	C	B	A	A	C	C	C	C	C	C	C	C	C	C	C	C	C	B	C	C	C	C	III
Organisation of tournaments	C	C	B	B	C	C	C	C	C	C	C	C	B	B	B	B	B	B	B	B	B	B	III

Environmental (risk) for normal operative situation (1) and abnormal operative situation or emergency (2)
A= high relevance
B= medium relevance
C= low relevance

Need for action: I= short-term, II= mid-term, III= long-term, IV= none

Fig. 4.19 Example of an environmental database for the ski area Bansko (AGL 2010)

What is more, environmental relevance is ascertained not only for the normal operative situation, but also for abnormal operative situations, i.e. in case of emergencies. For both aspects, "normal" and "emergency", environmental relevance is classified in three categories:

- A = high relevance
- B = medium relevance
- C = low relevance.

State of technology, qualification of personnel and regulations established by the business are considered in the operative situation.

A need for action is then summarised for each aspect, which can either indicate a short-term (I), mid-term (II), long-term (III) or no (IV) need for action.

In the following example from the ski area Bankso/Bulgaria (cf. Fig. 4.19), a particularly high relevance (A) is ascertained for summer piste management and renaturation due to the newly constructed ski area. A further important aspect in this example is snowmaking.

Checklist for evaluation of environmental aspects

- ☐ Has an evaluation been conducted to clearly identify the significant environmental aspects?
- ☐ Is spatial data adequately assessed and kept up to date?
- ☐ Are legal obligations involved in any of the aspects?
- ☐ Do the procedures and measures ensure constant and lasting compliance with environmental regulatory requirements and obligations?

4.6 Personnel Participation, Qualification, Training and Awareness

In this range of topics, one must differentiate between participation of employees during the first and subsequent environmental reviews and the qualifications and trainings that are necessary to minimise risk and environmental hazards. In the following, personnel participation during the first environmental review will be addressed first, followed by the long-term involvement of employees, their trainings and active participation in an environmentally compatible development of the business.

4.6.1 Personnel Participation During the Environmental Review

From the point of view of the regulation, personnel participation should ensure that employees feel actively involved and see themselves as driving forces for continued improvement and implementation of measures. From the point of view of business management, this should ensure that environmental performance is deeply rooted in the organisation and supported by all.

Since it is also important to involve employees as providers of information, the entire process should be begun under intensive personnel participation, involving all colleagues regardless of whether they sell lift tickets or are responsible for snowmaking. The EMAS regulation in particular underlines that personnel participation should involve all levels of the organisation.

In order to initiate this process during the course of the first environmental review, a kick-off event is recommended, where the most important points of implementing an environmental management system, as well as tasks, targets and roles of the employees are explained. If external experts are involved, they should be formally introduced during such an event. It is important that every employee knows whether, and which, information is required.

In a second step, once key figures for various aspects are already available, an analysis of strengths and weaknesses in the current situation is recommended (cf. Fig. 4.20).

The first environmental review plays a crucial role for continuous improvement of the environmental performance. Especially for ski areas, it constitutes the most important approach to this matter. Furthermore, measures should be taken to:

- define the business' internal organisation for the environmental review in detail in order to improve environmental performance,
- regulate and steer the flow of information on environmental issues (e.g. environmental panels, environmental representatives),
- promote cooperation between different departments (e.g. workgroups), and
- promote communication internally and externally.

Fig. 4.20 Personnel participation, example Schmittenhöhebahn (*Photo* Ulrike Pröbstl-Haider)

It is advisable to first enquire about the general opinion within the business, then to compare this with the analysis results. This approach benefits employees' motivation and promotes a lively discussion on the matters of possible environmental objectives and requirements for environmental management.

Emphasising positive aspects as well (strengths) increases acceptance for the process.

Ideally, the presentation should not only visualise technical issues but also evaluations as far as possible. If it should not be possible to provide or obtain complete data for individual map presentations or other aspects in time for the audit, this missing information should be emphasised in particular.

Since environmental management should be seen as a process, maximum transparency and traceability regarding assessment and evaluation methods should be pursued. A transparent process increases the willingness to participate. It is important to document and record intermediary steps and make them generally accessible.

Examples of appropriate forms of participation are the Metaplan technique during meetings, modified teamwork at a round table or the suggestion-book-system. The so-called sandwich method, a method with multiple components developed by Bischoff et al. (1996), has also proven useful in many cases. This method is suited whenever heterogeneous groups should work together or a large number of employees should be involved. The method should help reduce inhibitions to speak freely, thus allowing optimal incorporation of the experience and knowledge of employees as technical experts of the business (Table 4.20).

If top-level management is not present during the organised events for personnel participation, this gives employees greater freedom to express their opinions. Accord-

Table 4.20 Example for integration and motivation of employees using the sandwich method (Pröbstl et al. 2003: 85)

1	Introduction, welcoming words from business management
2	Presentation about tasks, thematic background and partial sections of the environmental management system as well as current challenges
3	Analysis of strengths and weaknesses by the employees using the Metaplan technique, individually or in groups
Break	
4	Summary of the Metaplan technique results and discussion of the communicated general opinions
5	Presentation by an external assessor about strengths and weaknesses
6	Discussion and joint elaboration of conformities and differences

Fig. 4.21 Map presentations are indispensable for the environmental review and derivation of measures (*Photo* Ulrike Pröbstl-Haider)

ingly, the top-level management should leave the meeting after giving the opening words. External moderation is also recommended for the same reason. Landscape-related analysis results should not only be shown in digital form, but should be represented clearly on maps and plans (cf. Fig. 4.21).

For ski areas with pronounced seasonality, choosing the right times for these talks is important. Late autumn and the period after the end of the season in spring are favourable. Respective planning of dates and the process in general facilitates a smooth execution of the environmental review.

A further meeting should concentrate on discussing targets, in order to emphasise how deficits will be approached. Participation in this step has the following aims:

- Personnel becomes involved and makes environmental management a personal issue.
- Employees of the business can bring forward their technical knowledge.
- Participants' thoughts and opinions can provide a test of plausibility of the measures' practicability, as well as of their consideration of sensible cost-benefit ratios.
- Participation helps agree on priorities and responsibilities.

Therefore, a draft of the catalogue of targets should already be on hand for this participation step. It facilitates the discussion with the participants.

Checklist for personnel participation

- □ Did the environmental review involve direct participation of personnel?
- □ Are there sufficient procedures and possibilities to exchange information and knowledge within the business? Are the tasks and responsibilities at different levels known and understood?
- □ Are there sufficient possibilities for training and professional development and is there a procedure for personnel qualification?
- □ Is the demand for further training assessed regularly?
- □ Are training measures for employees performing environmentally relevant tasks planned, conducted and documented systematically?
- □ Are new employees trained regularly with regard to the environmental management system?
- □ Are the environmental impacts of the business known?
- □ Are instruments established and procedures chosen that actively involve employees?

4.6.2 Raising Awareness, Training and Qualification

The first environmental review should be used as a basis for establishing an appropriate form of personnel participation in the long-term. The most important goal in this respect is to activate a lasting feedback of information, but also to receive and communicate success reports regarding implemented management measures. Awareness raising and internal communication are therefore ongoing tasks and challenges (also see Sect. 4.8 on communication).

Furthermore, the business must ensure that employees in sectors with a potential relevance for environmental impacts are sufficiently qualified, and that such experience is also documented and can be passed on. If need be, additional trainings should be provided for in this case. First, the need for training in connection with the environmental aspects and the environmental management system needs to be ascertained. Internal and external trainings therefore need to be assessed for different issues.

Internal communication and documentation ensures that all people working for the business

(a) recognise and implement the relationship between environmental policy and the associated procedures and requirements of the environmental management system;
(b) understand the significant environmental aspects and related actual or potential impacts connected to their activities within the business, and realise the environmental advantages that can be achieved through improved personal performance;
(c) understand their tasks and responsibilities in achieving the defined processes within the environmental management system, and
(d) can properly evaluate the potential consequences of diverging from defined procedures.

4.7 Objectives and Environmental Performance

4.7.1 Programme

Determining individual measures in a field of application as broadly diverse as the environmental management of ski areas is facilitated if a clear programme is developed or programmatic objectives are defined as a first step. The programme is subordinate to the generally formulated environmental policy and sets the frame for the elaboration of environmental objectives and targets, the appointing of responsible persons and defining of priorities. In the case of a follow-up review, the defined programme usually remains the same, but new targets are defined and updated with new figures where necessary.

The programme should be formulated in a way that it can actually be implemented with regard to the targets listed in the following. This means that the programme's emphasis should be on practical and measurable statements against the background of legal compliance.

The environmental programme should be presented to and enacted by the top management. Just like the definition of the environmental policy, the adoption of the necessary targets and measures is a matter for top-level management. Since the programme sets the frame for targets and measures for the employees, it should be as concrete and clearly understandable as possible.

Table 4.21 below shows different environmental objectives of ski areas and the associated targets (excerpts) and measures derived from the environmental policy.

Table 4.21 Presentation in table form of environmental objectives and measures (examples)

Requirements laid out in the environmental policy	Objective/Target	Measure
The ski area XXX is known for well prepared and maintained pistes that guarantee a successful winter season We aim at maintaining this situation with improvements in the fields of energy and water usage	Environmentally compatible snowmaking Minimising the demand for fuel, water and energy in preparation and snowmaking Quantifying resource usage	We consciously refrain from snowmaking after 10 March in order to avoid negative impacts on birds Installation of a reservoir lake should prevent potential impacts on natural bodies of flowing water and their ecosystems There are regular trainings for employees on efficient operation of the snowmaking systems under consideration of humidity and air temperature Resource usage is recorded internally, any exceedance over previous years is analysed and must be justified before business management
The ski area's location in valuable natural surroundings, with specific, often unique occurrences of habitats, animals and plants, commits us to sustainable and ecologically oriented management Development of the area should preserve this natural potential and—wherever compatible with use—enhance it Following re-greening of the piste surfaces, we will establish extensive sheep grazing in cooperation with local farms, thus promoting and stabilising species-rich alpine meadows	Improvement or restoration of natural meadows (in higher elevations) and near-natural meadows (in lower and middle elevations) Effective containment of erosion damages in the ski area Gradual improvement of the green vegetation coverage up to at least 50% coverage	Mulch-mowing is avoided. Conservation and development of diverse communities should be achieved through extensive use A programme of priorities is developed and implemented for individual piste sections concerning piste area drainage

4.7.2 *Targets, Measures and Realisation*

The individual targets are based, on the one hand, on the environmental policy and environmental programme, and on the other hand they should address issues

- with significant environmental potential
- with considerable potential for improvement
- with an urgent need for action or
- with a legal requirement for action.

Furthermore, the individual measures need to be clearly defined so that they can be planned, financed and implemented. Measureable, clearly defined parameters should be preferred. In defining individual targets, the regulation points out that the positions and interests of the local population and so-called interested parties also need to be considered. In some cases it might be necessary to give measures of particular significance for the general public a higher priority. Examples are noise abatement measures exceeding the legal requirements or voluntary measures to protect rare animal species.

In defining individual targets, potential use of local, regional or national environmental programmes or subsidies should also be assessed, for the benefit of the environment and the business. This could be the case for near-natural piste maintenance, for example, or for establishing infrastructure for renewable energy or use of biomass.

The regulation further points out that possibilities for technical improvement should be assessed when defining the targets. In ski areas, such possibilities are often found

- in piste maintenance and snow management in winter
- in piste maintenance in summer
- in resource usage and
- in transport and other indirect effects (traffic).

The following Table 4.22 connects individual measures with the persons responsible within the business, prioritisation and planned timing, which can be communicated internally and externally and ensures a high level of transparency.

As Table 4.22 shows, measures should be formulated with a view to their implementation. This means that business management shall ensure the availability of the resources required to establish, implement, maintain and improve the environmental management system. As illustrated, this also includes a clear regulation of responsibilities. Furthermore, the cost assessment should reckon in additionally required, specially trained personnel, the organisational infrastructure and technical aids.

The example above illustrates how tasks, responsibilities and competencies can be determined, documented and communicated to facilitate effective environmental management.

Table 4.22 Connecting individual measures, prioritisation and timeframe prioritisation

Target	Individual measure and area in question	Responsible employee	Timeframe
Piste maintenance and environmental protection			
Erosion prevention and piste re-greening Piste restoration Water discharge	Piste A and piste B	(Dpt. piste and snow management)	2nd quarter 20XX
Improving inventory of environmental data and measurement results	Setting up a database	Name (technical dpt.) and Name (dpt. piste and snow management)	4th quarter 20XX
Improving waste separation	Elaboration and implementation	Name (dpt. supplies and disposal, waste management)	3rd quarter 20XX
Promoting environmental know-how of employees	Concept for environmental training of employees	Name (dpt. finances and controlling)	2nd quarter 20XX
Information			
Improving customer information on environmental matters	Environmental information on the website	Name (dpt. advertising and marketing)	3rd quarter 20XX
Safety			
Improving safety	Training of employees on safety matters	Name (technical dpt.) and Name (dpt. piste and snow management)	2nd quarter 20XX
Avalanche protection	Implementation of approved solutions	Name (technical dpt.)	3rd quarter 20XX
Replacement of older lift systems for safety reasons	Replacement of the following cableways or tow lifts A, B, C	Name (technical dpt.)	4th quarter 20XX
Environmental management			
Environmental communication	Increased circulation of the environmental statement in brochure form	Name (dpt. advertising and marketing)	2nd quarter 20XX

4.7.3 Environmental Management, Operation and Personal Responsibility

According to ISO 14001 (as of 2015) top-level management of an organisation (e.g. CEO) must assign the responsibilities and competencies for relevant roles and make these known within the organisation. In the given case this would be the responsibility and competence for ensuring conformity of the environmental management system with the ISO 14001 standard; it would also involve reporting on such matters to the management level. Nomination of a special environmental management representative is thus no longer absolutely necessary, but is likely to remain common practice.

As shown in Table 4.22, it is important, therefore, to clearly define responsibilities and to assign them to individual departments.

The actual development of the environmental management system is the end of the whole process and involves, among other things, the realisation of the environmental policy with its environmental objectives. The overall management system incorporates the organisational structure, the responsibilities, procedures, processes and means for implementation.

The structure and size of the organisation determine to which degree of differentiation the environmental management is applied, and on which levels of the organisation.

Necessary allocations range from determining competencies, over implementing concrete individual measures, to communication and training. Furthermore, areas and issues need to be determined where monitoring, observation and supervision are important. Since cableway operations are usually small to medium-sized businesses, allocation of environmental issues to individual employees is particularly important.

In this regard, the following specific allocations to individual employees should be examined:

- Safety
- Fire safety
- Waste
- First aid
- Energy
- Environment.

Furthermore, threshold values for further legally required representatives, exceeding the focus competencies voluntarily allocated within the business, need to be examined.

Allocation of competencies and tasks within the business needs to be clearly documented to the outside.

4.8 External and Internal Communication and Documentation

As pointed out in the introduction, the external perception of environmental measures and their contribution to marketing are decisive from the point of view of the cableway sector. Therefore, communication of the process, planning (programme and objectives) and achieved implementation results are of particular importance.

Furthermore, as mentioned in Sect. 4.6, internal communication should be promoted. Every employee should be able to gain insight and thereby deliver the required performance.

A third area that needs to be considered in relation to communication, and which can partially be combined with the two aspects mentioned above, is documentation.

Regarding verifiable communication, the EMAS regulation states the following:

(1) Organisations shall be able to demonstrate an open dialogue with the public and other interested parties including local communities and customers with regard to the environmental impact of their activities, products and services in order to identify the public's and other interested parties' concerns.
(2) Openness, transparency and periodic provision of environmental information are key factors in differentiating EMAS from other schemes. These factors are also important for the organisation in building confidence with interested parties.
(3) EMAS provides flexibility to allow organisations to target relevant information to specific audiences while ensuring that all information is available to those who require it.

The environmental statement constitutes a crucial document providing an overview over external and internal communication. It summarises the entire process and its fundamental elements. The following Table 4.23 shows a typical structure of the environmental statement for a ski area and explains the fundamental points in a few key words.

With a properly structured environmental statement, the requirements regarding appropriate documentation are also fulfilled to a large extent. A further aspect of appropriate documentation is having documents, charts, tables and data compilations available for the first environmental statement, also in order to be able to update them continuously in digital form. The ski area information system in GIS and the documents mentioned above are required for the further planning, execution and monitoring of the processes making up the environmental management.

Suitable media also need to be determined for the dissemination of the internal and external documentation. Usually, the environmental statement is publicly available in printed form and as a download.

Comparing international approaches in the field of environmental communication, key environmental issues like energy, water, waste or biodiversity are often presented on separate pages, so that members of staff or external visitors are presented with thematically appealing pages. However, the environmental information should also

Table 4.23 Structure of an environmental statement (example)

Point in the structure of the environmental statement	Content in key words
Foreword	– Objectives – Naming of the type of certification – Relevance for internal and external communication
Characterisation of the business in figures and facts	– Characterisation of the business or organisation – Portrayal of the special characteristics of the natural environment, the pistes, recreation facilities, etc. – Size and structure of the space used for skiing, and demarcation of the area affected by the environmental management using map details
Systematic environmental management	– Responsibilities for environmental management – Involvement of external contributors and committees – Organisational diagram
Description of the environmental management system	– Indication of the underlying basis (e.g. EMAS regulation or ISO 14001:2015) – Tasks and composition of the prepared documents – Reference to environmental policy, programme, objectives and measures as basic structure – Reference to trainings, awareness raising and financial resources – Explanation of the environmental database and described risks – Environmental database for potential direct impacts – Reference to potential indirect environmental impacts
Environmental policy	– Basic principle – Scope of efforts – Delimitation – Environmental management system – Monitoring – Information
Direct environmental effects	– Snowmaking – Piste maintenance in winter – Safety – Piste maintenance in summer – Lift and cableway operation – Drinking water supply and waste water treatment – Waste management – Use of consumables and natural resources (electricity, water, petroleum products, emissions)
Indirect environmental effects	- Customer traffic on access roads and parking areas – Supply chain and partner businesses
Environmental programme	– Illustration in table form of targets/objectives, measures, responsibilities and realisation timeframes for relevant environmental issues, e.g. piste maintenance and environmental protection, safety, personnel training
Summary of available data on environmental performance	– Input-output analysis – Core indicators

(continued)

Table 4.23 (continued)

Point in the structure of the environmental statement	Content in key words
Reference to compliance with legal requirements, reference to environmental regulations	– Information on measurements and thresholds – Reference to significant legal regulations – Description of the legal database
Declaration of validity	– Name and licence number of the environmental verifier and date of validation
Information and feedback	– Contact persons within the business, additional information

be positioned within the website so that it is not only easy to find, but can also be included in the decision-making process of potential customers.

What is more, the regulation contains provisions for the handling of documents, i.e. their authorisation, updating, indication of revisions, etc., in order to prevent circulation of outdated or unauthorised documents (cf. Appendix, A 5.4). Furthermore, the exact type of documentation is to be determined for each of the fundamental issues with direct environmental relevance mentioned in the environmental statement (cf. Appendix, operational control). This also counts for potential indirect effects primarily due to the issues of traffic, supply firms, service providers and contractors.

Checklist for communication

- □ Is the language and presentation style concise and generally comprehensible?
- □ Is the data and information comprehensible and precise?
- □ Is the environmental statement available to the employees and the general public?
- □ Is the information easy to find on the website?

4.9 Documentation and Danger Prevention

4.9.1 Documentation

The environmental management systems EMAS and ISO 14001:2015 have certain requirements regarding documentation.[5] The issues that are of particular relevance for ski areas were already described in the individual technical chapters, supported by case examples. Aside from the environmental policy and objectives, the require-

[5]Some ski areas also recommend integrated management systems.

ments regarding documentation pertain to the documents that are necessary to ensure effective planning, execution and monitoring of environment-related measures.

In order to ensure efficient management, it needs to be determined within the business which content needs to be updated in which timeframes, and for which partial sectors new internal or official permits might be required. This might pertain, for example, to a limited permit for snowmaking or to regulatory obligations to provide snowmaking journals or documentation of snowmaking activity. The updating timeframes can differ considerably depending on usage or subject of protection. For example, seasonal records are required for water, waste and energy. In the case of vegetation, a sensible interval is 5 years (less, only in areas at risk of erosion). Furthermore, procedures must be determined for the questions of who is responsible for considering which documents, and how outdated documents should be archived.

Overall, documentation and proper handling of important records serve the purpose of being able to prove that the business complies with the requirements of ISO 14001 or the EMAS regulation. Furthermore, documentation proves that the desired results were in fact achieved. Therefore, companies must introduce and maintain procedures for identification, storage, retrieval and backup on the one hand, and a verifiable process for detecting, retracting and, if need be, eliminating outdated documents on the other hand.

It is essential to prevent unintentional use of outdated documents. An indication of which issues need to be sufficiently documented within the environmental management is provided by the evaluation of environmental aspects (see Sect. 4.5.2).

Examples for documents in ski areas are the environmental policy and the environmental programme. Examples for records are measurement records, maintenance reports (cableway systems) or the input-output analysis. These were already presented in the previous technical chapters with examples.

Compulsory documentation also pertains to interaction with suppliers, service providers or contractors. Since one of the aims is to achieve environmental improvements through these partners, any exercise of influence also needs to be documented. These comprehensive obligations regarding documentation also necessitate a targeted monitoring, i.e. regular supervision of the work processes that could have significant impact on the environment.

In ski areas, these usually include piste preparation including snowmaking, summer maintenance of ski pistes, water and electricity consumption and management of waste related to passenger transport. In all cases, appropriate records are required in order to prove that the work processes are in compliance with the environment-related objectives.

A good example is the recording of snowmaking, including the areas of application, date and time, weather conditions including wet-bulb temperature and the types of snowmaker used. This data, recorded by verifiable measurement devices, must be stored accordingly.

Obligations regarding documentation also include the issue of legal requirements. This is usually covered by the legal database (cf. Sect. 4.3) where status of verification and compliance are also noted. This regular evaluation of legal obligations must also be stored (cf. Sect. 4.5.1.3).

4.9.2 Preventive Measures, Emergency Preparedness and Danger Prevention

As illustrated at the beginning (Table 3.1), environmental management also contributes to a reduction of risk. The relevant measures in this case involve measures of correction and improvement as well as an emergency preparedness plan and targeted danger prevention measures.

The preventive measures involve developing procedures within the business that become effective when environmental impacts are caused by a deviation from "normal operation", e.g. piste preparation when there is an insufficient depth of snow in some sectors, operation of individual snowmakers at higher temperatures, etc. It is important, in these cases, to document which efforts are made to prevent potential problems, and which corrective actions were taken should a problem nonetheless have arisen, e.g. re-seeding in the section damaged by snow groomers.

Furthermore, in the case of danger prevention, a procedure must be introduced targeted at potential accidents. The organisation must demonstrate how it can react to potential emergency situations or accidents and how negative environmental impacts can be minimised or prevented.

If such emergencies have occurred, the determined measures should be reviewed and revised. Danger prevention can also necessitate inspections. In ski areas, this aspect should be examined regularly in relation to avalanches, preparation and cableway operation.

Checklist for documentation, handling of recording and procedures, and monitoring

- □ Does documentation encompass all documents that are important for effective planning, execution and monitoring of processes?
- □ Is observance of adequate updating timeframes ensured?
- □ Is it clear which employees/departments are responsible for updating, documenting and considering respective documents?
- □ Are procedures introduced and responsibilities defined for the most important environment-relevant aspects? Are identification, storage, retrieval and backup regulated?
- □ Has a procedure been introduced to identify outdated documents, retract them, archive or, if need be, eliminate them?
- □ Are courses of action documented for how important environmental aspects of suppliers and contractors are to be taken into account?
- □ Have provisions been made for potential emergency situations and incidents and have according procedures been determined?

References

Ammer U, Utschik H (1990) Entwurf einer Waldbiotopkartierung, wissenschaftliche Studie am Lehrstuhl für Landschaftstechnik, Munich (in German)

Arbeitsgruppe für Landnutzungsplanung (AGL) (2010) Auditing für das Skigebiet Bansko, Bulgarien - Dokumentation, Umweltpolitik und Umweltmanagementsystem, commissioned by the company Ulen AD, in cooperation with AVEGA, 161 pp (in German)

Birds Directive (2009) Directive 2009/147/EC of the European Parliament and of the Council of 30 November 2009 on the conservation of wild birds

Bischoff A, Selle K, Sinning H (1996) Informieren, Beteiligen, Kooperieren. Verlag Dorothea Rohn, Dortmund (in German)

Blaschke AP, Merz R, Parajka J, Salinas J, Blöschl G (2011) Auswirkungen des Klimawandels auf das Wasserdargebot von Grund - und Oberflächenwasser. Oesterr Wasser Abfallwirtsch 63(1–2):31–41 (in German)

Böhm R, Godina R, Nachtnebel HP, Pirker O (2008) Mögliche Klimafolgen für die Wasserwirtschaft in Österreichn: Bundesministerium für Land- und Forstwirtschaft, Umwelt und Wasserwirtschaft & Österreichischer Wasser- und Abfallwirtschaftsverband (eds), Auswirkungen des Klimawandels auf die österreichische Wasserwirtschaft, Vienna, pp 7–26 (in German)

Bundesministerium Für Verkehr, Innovation Und Technologie (bmvit) (2011) Erlass betreffend den Lawinenschutz im Bereich von Seilbahnen (Lawinenerlass 2011), Vienna, 8 pp (in German)

California Ski and Snowboard Safety Organization (CSSSO) (eds) (2009) Ski resorts celebrate national safety awareness week while keeping patrons in the dark about safety plans. Press release 16 (January 2009)

Conseil général de la savoie (2004) Inventaire et visualisation des câbles aériens dangereux pour les oiseaux en Savoie, Observatoire savoyard de l'environnement No. 12, pp 38–39 (in French)

Dambeck G, Wagner H (2007) Recht und Sicherheit im organisierten Skiraum, Oktober 2007. Reise- und Verlags-GmbH, Munich-Planegg, Interski Vermittlungs- (in German)

Dietmann T, Kohler U (2005) Skipistenuntersuchung Bayern: Landschaftsökologische Untersuchungen in den bayerischen Skigebieten. Bayerisches Landesamt für Umwelt, Augsburg (in German)

Ecosign (2005) Gutachten im Auftrag der Schmittenhöhebahn, Zell am See (in German)

Ellenberg H (1963) Grundlage der Vegetationsgliederung, 11. Die Vegetation Mitteleuropas, Stuttgart, Teil (in German)

Energy Efficiency Directive (2012) Directive 2012/27/EU of the European Parliament and of the Council of 25 October 2012 on energy efficiency, amending Directives 2009/125/EC and 2010/30/EU and repealing Directives 2004/8/EC and 2006/32/EC Text with EEA relevance

Energy Performance of Buildings Directive (2010) Directive 2010/31/EU of the European Parliament and of the Council of 19 May 2010 on the energy performance of buildings

European Union (2013) Umwelthaftungsrichtlinie - Schutz der natürlichen Ressourcen Europas, Amt für Veröffentlichungen der Europäischen Union, Luxemburg. (in German)

Formayer H (2013) Fachgutachten zur Einbeziehung des Klimawandels in das Umweltmangementsystem für das Skigebiet Schmittenhöhe in Zell am See. unveröffentlichtes Gutachten, Vienna. (in German)

Formayer H, Hofstätter M, Haas P (2007) Untersuchung der Schneesicherheit und der potenziellen Beschneiungszeiten in Schladming und Ramsau. Bundesministerium für Wissenschaft und Forschung im Rahmen von proVISION - Stratege, 43 pp (in German)

Kalz B, Knerr R, Brennenstuhl E, Kraatz U, Dürr T, Stein A (2015) Wirksamkeit von Vogelschutzmarkierungen an einer 380-kV-Freileitung im Nationalpark Unteres Odertal. NuL 47(4):109–116 (in German)

Keiler M, Sailer R, Jörg P, Weber C, Fuchs S, Zischg A, Sauermoser S (2006) Avalanche risk assessment—A multi-temporal approach, results from Galtür, Austria. Natural Hazards and Earth System Sciences 6:637–651

König A (2014) SiS-Prädikate: Prädikat "Geprüftes Skigebiet Deutschland", presentation from 09 May 2014 at the DSV Umweltbeirat, Rottach-Egern, 17 pp (in German)
Leicht H, Dietmann T, Kohler U (1993) Landschaftsökologische Untersuchungen in den Skigebieten des bayerischen Alpenraums – Darstellung und Methodik. In: Jahrbuch des Vereins zum Schutz der Bergwelt, Munich, pp 147–196 (in German)
Manhart M (2014) Personal communication, Lech
Manhart S, Manhart R (2015) Österreichisches Skirecht. Linde Verlag, Vienna (in German)
Mingo S (2000) Insuring Ski resorts: There's no business like snow business. Insurance Journal West Magazine. http://www.insurancejournal.com/magazines/features/2000/12/11/21207.htm. 29 November 2017
Oberstdorf/Kleinwalsertal Bergbahnen (2017) Modernes Energiemanagement von Skigebieten am Beispiel DAS HÖCHSTE – Fakten und Daten. https://www.ok-bergbahnen.com/unternehmen/natur-umwelt/modernes-energiemanagement.html. 29 November 2017 (in German)
Österreichischer Wasser- und Abfallwirtschaftsverband (ÖWAV) (Hrsg.) (2014) Klimawandelauswirkungen und Anpassungsstrategien in der österreichischen Wasserwirtschaft, Vienna, 10 pp (in German)
Plachter H (1989) Zur biologischen Schnellansprache und Bewertung von Gebieten, Schriftenreihe für Landschaftspflege und Naturschutz 29, pp 107–135 (in German)
Pröbstl-Haider U, (2010) Unterlagen für den Umweltaudit für das Skigebiet Bansko, Bulgarien, unveröffentlichtes Fachgutachten, Polling
Pröbst-Haider U, Dorsch C (2013) Auditing für das Skigebiet Schmitten in Zell am See, zusammenfassender Bericht, 179 pp (in German)
Pröbstl U (1990) Skisport und Vegetation, Die Auswirkungen des Skisports auf die Vegetation der Skipiste, vol 2. Stöppel-Verlag KG, Weilheim (in German)
Pröbstl U (2001) Skigebiete in den Bayerischen Alpen. Ergebnisse einer ökologischen Studie. DSV-Umweltreihe, Band 7. Stöppel-Verlag KG, Weilheim. (in German)
Pröbstl U (2006) Kunstschnee und Umwelt - Entwicklung und Auswirkungen der technischen Beschneiung. Haupt Verlag, Bern Stuttgart Vienna (in German)
Pröbstl U, Roth R, Schlegel H, Straub R (2003) Auditing in Skigebieten. Freiburg, Vaduz (in German)
Renat (2000) Audit von Skigebieten: Avifaunistisch-ökologische Beurteilung. Zwischenbericht. Unpubished, Schaan (in German)
Schmid H, Doppler W, Heynen D, Rössler M (2012) Vogelfreundliches Bauen mit Glas und Licht, 2, überarbeitete edn. Schweizerische Vogelwarte, Sempach (in German)
Schmittenhöhebahn AG (2014) Unterlagen zum Audit Schmittenhöhebahn, Zell am See (in German)
Schuster L (2016) Effizientes Schneemanagement durch den Einsatz von Schneehöhenmesssystemen. Abschlussarbeit an der Fachhochschule Vorarlberg, Dornbirn (in German)
Seilbahnen Schweiz (2011) Checkliste Verkehrssicherungspflicht für Sommeraktivitäten, Arbeitsgruppe Sommeraktivitäten von Seilbahnen Schweiz, Bern, 38 pp. (in German)
Stabentheiner J (2016) Pistensicherung und verwandte Fragenkreise - 35 Jahre Seilbahnsymposium, ZVR 2016/104 (in German)
User MB, Erz W (Hrsg.) (1994) Erfassen und Bewerten im Naturschutz, Wiesbaden (in German)
Voets C (2009) Welche Auswirkungen hat das neue Umweltschadensgesetz auf Planungs- und Umweltprüfinstrumente wie z.B. FFH-VP und UVP und welchen Beitrag können diese Instrumente leisten, Umweltschäden zu vermeiden? UVP Report 23, No. 5, 2009, pp 240–244. (in German)
Zeitler A (2001) Veränderungen des winterlichen Raum-Zeit-Musters von Rauhfußhuhn-Arten durch Skifahrer und die Begrenzung ihrer Folgen. Laufener Seminarbeiträge 1/01, pp 31–35, Laufen/Salzach (in German)
Zeitler A (2014) Vögel um Skistationen, Fachgutachten, Immenstadt, 24 pp (in German)

Chapter 5
Management Review and Internal Audit

Abstract This chapter explains the internal audit and its recommended content. During this step, top management shall review the environmental management system in defined intervals and identify potential for improvement as well as possible need for changes. This process is crucial for the continuing suitability, adequacy and effectiveness of the system and ensures continual improvement.

Keywords Internal review · Management evaluation · Changes of the management system · Objectives and targets

A further element of the regular internal reviews is the so-called management evaluation or management review. During this step, top management shall review the environmental management system in defined intervals, to ensure its continuing suitability, adequacy and effectiveness. Reviews shall include an evaluation of potential for improvement and the need for changes to the environmental management system, including the environmental policy and environmental objectives and targets.

Records of the management reviews must be kept. The following aspects are to be considered during the course of the management review:

- results of internal audits and evaluations of compliance with legal requirements and with other requirements to which the organisation subscribes;
- comments from external interested parties, including complaints;
- the environmental performance of the organisation;
- the extent to which objectives and targets have been met;
- status of corrective and preventive actions;
- follow-up actions from previous management reviews;
- changing circumstances, including developments in legal and other requirements related to the business' environmental aspects; and
- recommendations for improvement.

The results of the management reviews shall include any decisions and actions related to possible changes to environmental policy, objectives, targets and other elements of the environmental management system, consistent with the commitment to continual improvement.

U. Pröbstl-Haider et al., *Environmental Management in Ski Areas*, SpringerBriefs in Environmental Science, https://doi.org/10.1007/978-3-319-75061-3_5

Chapter 6
Verification and Validation Through Environmental Verifiers

Abstract The verification and validation process under the EMAS regulation demand a specific procedure explained in this chapter. This step examines whether the requirements of the EMAS regulation are fulfilled, which may only be conducted by authorised external environmental verifiers. The description includes the necessary qualifications of such verifiers as well as the various aspects that need to be examined by them.

Keywords Verification process · Validation · National licensing · Accreditation

The verification and validation process may start if the following aspects have been terminated:

- a functioning environmental management system has been implemented,
- an environmental review has been conducted and documented,
- the management review and internal audit have been completed and
- measures have been defined and an environmental statement completed in the ski area. Overall the external verification can be sought as a conclusion to the whole process.

This verification may only be conducted by accredited or specially authorised environmental verifiers.[1] A list of authorised environmental verifiers is available at each of the national licensing or accreditation bodies. The environmental verifier must further be accredited for the relevant so-called NACE code. In the case of ski areas, the most relevant code in question is: H 49.39-1 (operation of chair and tow lifts). If the ski area also includes other facilities, additional NACE codes might also be relevant: R 93.11 (operation of sports facilities, e.g. winter sports arenas and stadiums) or tourism facilities such as inns, hotels and guesthouses, summarised under I 55.10.

[1]The responsible environmental authorities in each respective country are to be contacted. Only accredited or authorised environmental verifiers may carry out this function, provided they are accredited for the NACE code relevant to the activities of the organisation. The list of environmental verifiers accredited in Austria is maintained by the Federal Ministry of Agriculture, Forestry, Environment and Water Management: http://www.lebensministerium.at/umwelt/betriebl_umweltschutz_uvp/emas/Umweltgutachter/Umweltgutachterliste.html.

U. Pröbstl-Haider et al., *Environmental Management in Ski Areas*, SpringerBriefs in Environmental Science, https://doi.org/10.1007/978-3-319-75061-3_6

In principle, environmental verifiers can also be chosen from other member states than the organisation's own member state. It must be considered, however, that verification and validation also involve legal questions to a considerable extent. Therefore, at least in this regard, such environmental verifiers will often require assistance from national colleagues.

The **verification** process examines whether the environmental review, environmental policy, environmental management system and internal environmental audit and its implementation fulfil the requirements of the EMAS regulation.

Through **validation**, the environmental verifier confirms that the information and data in the environmental statement and its updates are reliable, credible and correct and meet the requirements of the EMAS regulation.

In detail, the following aspects are examined by the environmental verifiers:

1. The environmental verifiers examine whether the ski area complies with all the requirements of the EMAS regulation with respect to the environmental review, environmental management system, environmental audit and its results, and the environmental statement.
2. They examine whether the ski area complies with all relevant Community, national, regional and local environmental regulations and does not violate applicable environmental law. If the organisation can prove that it has taken timely measures to re-ensure compliance with legal requirements, the environmental verifier can sign the declaration on verification and validation activities.
3. Furthermore, the environmental verifiers examine the continued improvement of the ski area's environmental performance.
4. They analyse the reliability, credibility and correctness of the data used in in the EMAS environmental statement, as well as of all environmental information to be validated.

The environmental verifiers generally conduct an on-site EMAS Audit once a year. In the case of small organisations, a derogation according to Article 7 of the EMAS regulation might be applicable.

Chapter 7
Registration Process (EMAS)

Abstract This chapter explains the last steps of the EMAS process, its registration by the respective environmental agency and lists the required documents. In addition, the requirements for organisations to renew their EMAS registration are addressed. Austria and the documents provided by its national Environment Agency serve as an example to describe the procedure.

Keywords Registration · Environmental agency · Validated environmental statement

As soon as the environmental management system has been introduced and verified and the EMAS environmental statement has been validated, the ski area or the organisation applies for registration to the competent body. The competent body for EMAS registrations in Austria is the Environment Agency Austria (Umweltbundesamt).[1]

The following documents must be submitted with the initial application[2]:

- the validated environmental statement (in electronic or printed form);
- the signed declaration of the environmental verifier;
- the completed application form with details on the organisation/ski area, its locations and the environmental verifier.

After the Environment Agency Austria has contacted the competent authorities and satisfied itself that there are no violations against applicable environmental regulations, the registration fee is charged. After payment of this fee, the ski area is added to the EMAS register.

[1]Information on EMAS by the Environment Agency Austria: http://www.umweltbundesamt.at/emas.

[2]Form for initial registration or registration renewal of EMAS organisations by the Environment Agency Austria: https://www.bmlfuw.gv.at/umwelt/betriebl_umweltschutz_uvp/emas/ansprechpartner-und-formulare/FormulareEMAS.html.

U. Pröbstl-Haider et al., *Environmental Management in Ski Areas*, SpringerBriefs in Environmental Science, https://doi.org/10.1007/978-3-319-75061-3_7

In order to **renew EMAS registration**, every three years a registered ski area must:

- have the full environmental management system and programme for environmental audits and its implementation verified;
- prepare an environmental statement and have it validated by an environmental verifier, and
- provide the validated environmental statement and completed form.

In the interim years, the registered ski area must fulfil the following requirements:

- Conduct an assessment of environmental performance and compliance with applicable environmental regulations (internal and external audit);
- Update the environmental statement and have it validated by an environmental verifier, and
- Provide the validated, updated environmental statement to the Environment Agency.

Appendix

Annex IV of the Habitats Directive for Consideration of Legal Matters Relevant to Alpine Ski Areas in Germany and Austria

Based on information of the Environment Agency Austria (Umweltbundesamt)[1] and the Bavarian Environment Agency (Bayerisches Landesamt für Umwelt),[2] those species from Annex IV of the Habitats Directive are listed in the following, which occur in the German and Austrian Alpine region and which must therefore receive particular consideration in planning and management of ski areas, with regard to European species conservation legislation and the Environmental Liability Directive (priority species whose preservation is of particular significance are indicated by an asterisk). This list also provides a reference point for consideration of the issue in other member states of the European Union.

Species name		Bavarian Alpine region			Austrian Alpine region
Scientific	English	D 68	D 67	D 66	
Amphibians					
Bombina variegata	Yellow-bellied toad	x	x	x	x
Bufo calamita	Natterjack toad		x	x	x
Bufo viridis	European green toad		x	x	x
Hyla arborea	European tree frog	x	x	x	x
Rana arvalis	Moor frog			x	x
Rana dalmatina	Agile frog	x	x	x	x
Rana lessonae	Pool frog	x	x	x	x
Salamandra atra	Alpine salamander	x	x	x	x

(continued)

[1]Ellmauer 2016.
[2]http://www.lfu.bayern.de/natur/sap/arteninformationen/.

U. Pröbstl-Haider et al., *Environmental Management in Ski Areas*, SpringerBriefs in Environmental Science, https://doi.org/10.1007/978-3-319-75061-3

(continued)

Species name		Bavarian Alpine region			Austrian Alpine region
Scientific	English	D 68	D 67	D 66	
Triturus carnifex	Italian crested newt				x
Triturus cristatus	Northern crested newt	x	x	x	x
Bats					
Barbastella barbastellus	Barbastelle	x	x	x	x
Eptesicus nilssonii	Northern bat	x	x	x	x
Eptesicus serotinus	Serotine bat	x	x	x	x
Hypsugo savii	Savi's pipistrelle				x
Miniopterus schreibersii	Common bent-wing bat				x
Myotis alcathoe	Alcathoe bat			x	
Myotis bechsteinii	Bechstein's bat	x	x	x	x
Myotis blythii	Lesser mouse-eared bat				x
Myotis brandtii	Brandt's bat	x	x	x	x
Myotis daubentonii	Daubenton's bat	x	x	x	x
Myotis emarginatus	Geoffroy's bat	x	x	x	x
Myotis myotis	Greater mouse-eared bat	x	x	x	x
Myotis mystacinus	Whiskered bat	x	x	x	x
Myotis nattereri	Natterer's bat	x	x	x	x
Nyctalus leisleri	Lesser noctule	x	x	x	x
Nyctalus noctula	Common noctule	x	x	x	x
Pipistrellus kuhlii	Kuhl's pipistrelle			x	x
Pipistrellus nathusii	Nathusius' pipistrelle	x	x	x	x
Pipistrellus pipistrellus	Common pipistrelle	x	x	x	x
Pipistrellus pygmaeus	Soprano pipistrelle	x	x	x	x
Plecotus auritus	Brown long-eared bat	x	x	x	x
Plecotus austriacus	Grey long-eared bat	x	x	x	x
Plecotus macrobullaris	Alpine long-eared bat				x
Rhinolophus ferrumequinum	Greater horseshoe bat			x	x
Rhinolophus hipposideros	Lesser horseshoe bat	x	x	x	x
Vespertilio murinus	Parti-coloured bat	x	x	x	x

(continued)

(continued)

Species name		Bavarian Alpine region			Austrian Alpine region
Scientific	English	D 68	D 67	D 66	
Orthoptera (Grasshoppers)					
Isophya costata	Keeled plump bush-cricket				x
Paracaloptenus caloptenoides	No common name				x
Saga pedo	Predatory bush cricket				x
Beetles					
Buprestis splendens	Goldstreifiger				x
Carabus (variolosus) nodulosus	No common name			x	x
Cerambyx cerdo	Great capricorn beetle				x
Cucujus cinnaberinus	No common name	x	x	x	x
Dytiscus latissimus	No common name			x	
Graphoderus bilineatus	No common name			x	x
*Osmoderma eremita**	Hermit beetle			x	x
Phryganophilus ruficollis	No common name				x
*Rosalia alpina**	Rosalia longicorn	x	x	x	x
Odonata (Dragonflies)					
Cordulegaster heros	Balkan goldenring				x
Leucorrhinia albifrons	Dark whiteface		x	x	x
Leucorrhinia caudalis	Lilypad whiteface			x	x
Leucorrhinia pectoralis	Large white-faced darter			x	x
Ophiogomphus cecilia	Green snaketail			x	x
Stylurus flavipes (Gomphus flavipes)	River clubtail			x	
Sympecma braueri	Siberian winter damsel		x	x	x
Bivalvia					
Unio crassus	Thick-shelled river mussel			x	x
Rodents					
Castor fiber	Eurasian beaver	x	x	x	x
Dryomys nitedula	Forest dormouse	x	x		x
Muscardinus avellanarius	Hazel dormouse	x	x	x	x
Sicista betulina	Northern birch mouse	x			x
Spermophilus citellus	European ground squirrel				x

(continued)

(continued)

Species name		Bavarian Alpine region			Austrian Alpine region
Scientific	English	D 68	D 67	D 66	
Carnivorans					
Lutra lutra	Eurasian otter	x	x	x	x
Lynx lynx	Eurasian lynx	x			x
*Ursus arctos**	Brown bear				x
Reptiles					
Coronella austriaca	Smooth snake	x	x	x	x
Elaphe longissima	Aesculapian snake	x		x	x
Emys orbicularis	European pond turtle			x	
Lacerta agilis	Sand lizard	x	x	x	x
Lacerta horvathi	Horvath's rock lizard				x
Lacerta viridis	European green lizard				x
Natrix tessellata	Dice snake				x
Podarcis muralis	Common wall lizard		x	x	x
Vipera ammodytes	Nose-horned viper				x
Lepidoptera (Butterflies)					
Coenonympha hero	Scarce heath		x	x	x
Coenonympha oedippus	False ringlet				x
Erebia calcaria	Lorkovic's brassy ringlet				x
Hypodryas maturna	Scarce fritillary	x		x	x
Leptidia morsei	Fenton's wood white				x
Lopinga achine	Woodland brown	x	x	x	x
Lycaena dispar	Large copper				x
Lycaena helle	Violet copper		x	x	x
Maculinea arion	Large blue	x	x	x	x
Maculinea nausithous	Dusky large blue	x	x	x	x
Maculinea teleius	Scarce large blue	x	x	x	x
Parnassius apollo	Apollo	x	x	x	x
Paranassius mnemosyne	Clouded Apollo	x	x		x
Proserpinus proserpina	Willowherb hawkmoth				x
Gastropods (Snails)					
Anisus vorticulus	Lesser ramshorn snail			x	x
Theodoxus transversalis	Striped nerite			x	

(continued)

(continued)

Species name		Bavarian Alpine region			Austrian Alpine region
Scientific	English	D 68	D 67	D 66	
Vascular Plants					
Adenophora liliifolia	Ladybells				x
Apium repens	Creeping marshwort	x	x	x	x
Aquilegia alpina	Alpine columbine				x
Asplenium adulterinum	Ladder spleenwort				x
Botrychium simplex	Least moonwort				x
Campanula zoysii	Zois' bellflower				x
Cypripedium calceolus	Lady's slipper orchid	x	x	x	x
Dracocephalum austriacum	Pontic dragonhead				x
Eryngium alpinum	Alpine sea holly				x
Gladiolus palustris	Marsh gladiolus	x	x	x	x
Himantoglossum adriaticum	Adriatic lizard orchid				x
Ligularia sibirica	No common name				x
Lindernia procumbens	No common name				x
Liparis loeselii	Fen orchid	x	x	x	x
Myosotis rehsteineri	No common name			x	x
Physoplexis comosa	Tufted horned rampion				x
Pulsatilla grandis	Greater pasque flower				x
Rhododendron luteum	Yellow azalea				x
Saxifraga hirculus	Yellow marsh saxifrage			x	
Serratula lycopifolia	No common name				x
Spiranthes aestivalis	Summer lady's-tresses	x	x	x	x
*Stipa styriaca**	No common name				x
Trifolium saxatile	No common name				x

D 68: Northern Limestone Alps
D 67: Swabian-Bavarian Pre-Alps
D 66: Bavarian Alpine Foreland

Furthermore, according to the Birds Directive (Directive 2009/147/EC), all species of wild birds inhabiting the European area must be considered.

Reference

Birds Directive, Directive 2009/147/EC of the European Parliament and of the Council of 30 November 2009 on the conservation of wild birds

Bibliography

Ellmauer T (Environment Agency Austria/Umweltbundesamt), E-Mail from 08 June 2016

Environmental Liability Directive, Directive 2004/35/CE of the European Parliament and of the Council of 21 April 2004 on environmental liability with regard to the prevention and remedying of environmental damage

Habitats Directive, Council Directive 92/43/EEC of 21 May 1992 on the conservation of natural habitats and of wild fauna and flora

http://www.emas.de/fileadmin/user_upload/06_service/PDF-Dateien/UGA_Infoblatt_EMAS.pdf (29 November 2017)

Renewable Energy Directive, Directive 2009/28/EC of the European Parliament and of the Council of 23 April 2009 on the promotion of the use of energy from renewable sources and amending and subsequently repealing Directives 2001/77/EC and 2003/30/EC (Text with EEA relevance)

MIX
Papier aus verantwortungsvollen Quellen
Paper from responsible sources
FSC® C105338

If you have any concerns about our products,
you can contact us on
ProductSafety@springernature.com

In case Publisher is established outside the EU,
the EU authorized representative is:
Springer Nature Customer Service Center GmbH
Europaplatz 3, 69115 Heidelberg, Germany

Printed by Libri Plureos GmbH
in Hamburg, Germany